Who's Watching You?

Who's Watching You?

An Exploration of the Bigfoot Phenomenon in the Pacific Northwest

Linda Coil Suchy

Editing and Consulting: Christopher L. Murphy

ISBN 978-0-88839-664-8 [print]
ISBN 978-0-88839-674-7 [e-book]

Second printing 2010

Cataloging in Publication Data

Coil Suchy, Linda, 1951–
Who's watching you? : an exploration of the bigfoot phenomenon in the Pacific northwest / Linda Coil Suchy ; editing and consulting: Christopher L. Murphy.

Also available in electronic format.
Includes bibliographical references and index.
ISBN 978-0-88839-664-8

1. Sasquatch. I. Murphy, Christopher L. (Christopher Leo), 1941– II. Title.

QL89.2.S2C63 2009 001.944 C2009-902036-X

Printed in China — Fpoint International Developments Ltd.

Editor: Chris Murphy
Production: Chris Murphy
Cover Design: Ingrid Luters
Photograph and illustration copyrights and credits: See page 395

Published simultaneously in Canada and the United States by

HANCOCK HOUSE PUBLISHERS LTD.
19313 Zero Avenue, Surrey, B.C. Canada V3S 9R9
(604) 538-1114 Fax (604) 538-2262

HANCOCK HOUSE PUBLISHERS
1431 Harrison Avenue, Blaine, WA U.S.A. 98230-5005
(604) 538-1114 Fax (604) 538-2262

***Website:* www.hancockhouse.com**
***Email:* sales@hancockhouse.com**

Dedication

To my beloved grandmother, who opened my eyes to a world of beauty, mystery, and intrigue. A day does not pass that I do not think of her. God bless you Gram.

And also to my daughter Amanda (beautiful, charming, and awe-inspiring); my son Shane (multi-talented, strong, my rock); my grandsons—Tristian (my love, my hero, my light), Sam (a gift from god), and Kyro (who's perfect in every way); my granddaughters—Karrie Ann (beautiful artist with the smiling eyes) and Audrey Jane (a pretty princess who makes me laugh).

And to my grandfather Jerry (the love of Gram's life), and to my lifelong friend Bubs (my sister and partner in crime), and my friend Bill White (the wind beneath my wings)—all whose unfailing support, love, and encouragement got me through the "rough times."

A huge thank you goes out to all of the eyewitnesses for their contributions of great reports. Your names are too many to list, but I know who you are, each and every one.

And finally, I want to acknowledge and give a very special thank you to all of our veterans who served so bravely for us. We stand proud and salute you. Thank you all. God Bless America.

Dedication

Contents

About the Author and Editor

Linda Coil Suchy is a bigfoot researcher and writer. She's led a full life, her background consisting of many roles and adventures, including, law enforcement, shuttle driver, and heavy construction dump truck driver (frequently writing every chance she got during downtime). She is now a professional retiree with plenty of time to write and play with the grandkids. She resides in Everett, Washington, along with her little dog Oliver and fat cat Gracie.

Christopher L. Murphy retired in 1994 after thirty-six years of service with the British Columbia Telephone Company (now Telus). Chris got involved in the sasquatch mystery when he met René Dahinden in 1993. In 1996, Chris republished Roger Patterson's 1966 book, *Do Abominable Snowmen of America Really Exist?* and Fred Beck's book, *I Fought the Apemen of Mt. St. Helens.* He then went on to author *Bigfoot in Ohio: Encounters with the Grassman,* with Joedy Cook and George Clappison of Ohio.

In 2000, Chris embarked on a project to assemble a comprehensive pictorial presentation on the sasquatch. This initiative led to the 2004 sasquatch exhibit at the Vancouver (BC) Museum and the publication of his book *Meet the Sasquatch,* written in association with John Green and Thomas Steenburg. In due course, he wrote a supplement for Patterson's book and updated his Ohio book. Both were published by Hancock House Publishers.

In June 2005, Chris' sasquatch exhibit traveled to the Museum of Mysteries in Seattle, then in June of the following year, it opened at the Museum of Natural History in Pocatello, Idaho, where it was shown for fifteen months.

Chris has also attended and presented at many sasquatch symposiums, and has taken part in several television documentaries on the subject.

Acknowledgments

I wish to thank the many researchers and others who gave freely of their time and knowledge in all aspects of this work:

Dmitri Bayanov, S.D.Baker, Tom Biscardi, Peter Byrne, Ray Crowe, Tom Cusino, Dr. Henner Fahrenbach, Sean Fries, Bob Gimlin, John Green, Dr. Jeffrey Meldrum, Robert W. Morgan, John Pickering, Ray Rosa, Michael Rugg, and Thomas Steenburg.

Introduction

> Mystery creates wonder, and wonder is the basis of man's desire to understand.
>
> —Neil Armstrong

I first became interested in bigfoot when as a young girl my grandmother told me of visits by the creature to her home (and extensive vegetable garden) in Bellingham, Washington. (I provide full details of this experience in the first chapter.) From that point on, my fascination for the creature continued, and I began to read and collect everything I could find on the subject.

I have been interviewing and documenting eyewitness accounts for many years now, and the more I inquired, the more I opened doors for people to relate their own personal experiences.

In 2003, I sat down to write my Gram's story for my grandchildren. While writing of her experiences, I felt an urge to expand beyond just this account and started to contemplate a book on the bigfoot subject.

As I turned things over in my mind, the thought occurred to me that I should find out what people in general might want to see in such a book. I therefore put out a call over the Internet specifically asking people this question. In other words, what burning questions would they like to have answered? I also invited those who have had bigfoot-related experiences to please come forward and tell me about them.

The response was astounding—over 8,000 people replied with comments, questions, and hundreds of eyewitness accounts, some of which were highly intriguing, and these I added to my special collection for possible publication.

I then approached various professionals and seasoned researchers with the aim of getting answers to the questions people had asked. Here again the response was excellent; it totally exceeding my expectations. I prepared and sent out interviews to which I received highly detailed and in-depth replies that fully addressed every issue.

To round off what I now saw as a highly comprehensive discussion on the subject, I researched and included the "classics" in bigfoot history, together with many other stories on record with the International Bigfoot Society and other sources.

As I progressed, I recognized a need to document some of the sasquatch/bigfoot attractions in the west for the benefit of travelers who might wish to experience a little of the bigfoot phenomenon firsthand. To this end, I researched and provided profiles on many attractions, including full directions, where necessary, on how to find them.

During this time, I contacted Chris Murphy as to information on bigfoot/sasquatch attractions. This led to sending my book manuscript to Hancock House Publishers and Chris, who voluntarily reviews books of this nature for Hancock. Chris said that my work had good publishing possibilities, and then agreed to provide consultation and editing, along with providing (or arranging to provide) many more images. Chris recommended the work for publishing by Hancock House.

That, dear reader, is how this work came to be, and I sincerely hope you enjoy reading it as much as I enjoyed writing it and working with the many wonderful people in the "bigfoot fraternity."

—LCS

Chapter 1

Gram's Sasquatch Story

If all the beasts were gone, men would die from a great loneliness of spirit. For whatever happens to the beasts, soon happens to man. All things are connected.

—Chief Seattle

The story you are about to read is very personal and special to me. It is a story I grew up with which brought me much wonderment and awe throughout my life. With my early retirement came the time and the need to tell of my grandmother's experience with a sasquatch. When this time finally presented itself, it fulfilled a long-time desire to document in written word her story, for my grandchildren as well as for those across the great divide that care to hear, and know. I now present my grandmother's story, followed by a large collection of most magnificent and personal accounts from others, for your enlightenment, enjoyment, excitement, contemplation, and skepticism as well. I hope that you enjoy.

—Linda Coil Suchy

Grandma moved to Washington State in 1959. Gram, as I called her, and her little dog Timmy, took up housekeeping in a cozy little house in the country. She was surrounded by woods and close to a river. It was that very first summer, "That's about the time things came up missing," she said. The first thing she noticed was the missing blanket, "stolen right off my clothesline!" The next week it was two blankets. She stopped leaving them out at night after that. "They never took my clothes, not even one pair of undies thank goodness, it was just the blankets they were after."

Gram had a huge vegetable garden and grew perfect and colorful foods. Tomatoes, potatoes, carrots, green beans, lettuce, cabbage, cucumbers, squash, pumpkins, raspberries, strawberries, and other vegetables as well. There was even a grape harbor under which she had a picnic table and chairs.

Gram with a bouquet of beet
Her vegetable garden was h
little world, and we all enjoye
the fruits of her "labor of love

Shortly after the blankets came up missing, she began to notice fruits and vegetables missing from the garden. Of course there were problems with the occasional deer, raccoons, and a rabbit or two, but this was different. The vegetables such as the tomatoes appeared to have been neatly plucked from their vines and whole squashes and zucchinis came up missing. Now if a deer or a rabbit were to eat from a squash they would nibble at it, or eat a hole out of it. Whoever or whatever this was picked and took the whole squash, and Gram's squashes grew huge and heavy.

One day while working in the garden, she noticed tracks in the soft dirt around the potato patch. As she took a closer look she saw they were not deer tracks, they were not raccoon prints, nor were they claw prints like a bear would leave. In fact, these were not any kind of paw, claw, or hoof prints. They had five toes! These were footprints, huge footprints! She thought to herself, "If this is one of the locals, they sure have some big feet!" She estimated the prints to be about eighteen inches long.

That evening when darkness fell, she pulled her favorite chair up to the window, and with Timmy on her lap she sat in the dark and waited. Around midnight she heard a howl, a loud howl like nothing she had ever heard before. Timmy shook and whined.

It was not a coyote, she knew that. And it was not a cougar. She heard it repeat several times, then the howling was followed by several hoots. She used to perform these hoots for me, "Whoop! Whoop!" she would go. The howling she was never able to imitate. She never saw anything that night, just heard the sounds of something very strange.

The next night she was too tired to stay up, so she slept with one ear alert and never heard a thing.

The third night she left a giant squash, two tomatoes and a basket of string beans outside on the picnic table, and she stayed up again watching out her window, this time with the rifle standing next to her. And this night she wasn't disappointed.

As Gram sat with Timmy in the dark, she found herself lost in thoughts of the past. Suddenly Timmy's head jerked and his little ears stood alert as he began to tremble. Off her lap he jumped, running straight under the bed. Gram never heard a thing even though her window was open a crack. She stood, looked out toward her garden, and there standing in front of her raspberry bush was a huge creature. "A dark brown, hairy beast," is how she described it, "at least eight feet tall. He would have to duck just to come inside the house." She watched as it ate the berries, grabbing handfuls and shoving them into its mouth. "I thought it was a big gorilla escaped from the zoo or the circus, but I knew it wasn't. This was no gorilla!"

When finished with the berries, he wondered over to the tomatoes and she watched him very gently pluck one, hold it up to inspect it, then shove the whole thing into his mouth. He then took the squash from the picnic table, and with giant strides and one arm swinging he walked away into the woods.

She used to imitate his walk for me and I would laugh (her walk being very similar to Groucho Marx). She would say, "I'm not being funny, that's how he walks!" I asked her once what she felt that first night and mentioned that she must have been terrified. Her answer was, "I just thought, why did he pluck a tomato off the vine when there was two perfectly good ones on the table and all he took was the squash!"

Gram wasn't one to be afraid of much of anything. She's the one you could depend on to save you from the big spider or the one to take her shoe (if not her hand) and smash the cockroach. I remember once as a young child living in California, we had a man try to break into our house through a narrow bathroom window. Gram was so mad she grabbed a huge butcher knife on her way outside. The man was halfway through the window; she pulled him back out by his ankles and chased him all the way up the alley. We never saw him again!

She never waited up for the creature again; from that day forward she would leave something on the picnic table for him. "He is some unfortunate creature who is just hungry, that's all. And I share and share alike." She said he only came in the summer and fall months, and he came year after year. She found that he preferred the squash and pumpkins, so that very next summer she began to make cakes out of his favorite vegetables to leave on the table for him. About once a week she would bake him one in a Bundt cake pan. She said he was such a big guy and his hands were so huge that he could pick the cake up by the hole in the center and eat it like a donut.

He must have loved those cakes, because soon after he began to leave her little tokens of appreciation on the picnic table. There would be an apple, a pine cone, a flower, or a pretty rock. She kept those gifts on the window sill in her kitchen.

I remember an apple he once left for her, which had a huge worm hole in it. She showed it to me and said, "Look, there's a fat worm living in this apple. I don't think he cares and eats the worm and all." It seems to me that apple sat up on the window sill for a very long time. I've never seen an apple last so long.

Every summer she would also leave a blanket out on the clothesline, but it was never taken. Nor was the garden ever raided again; only the cake and the fruit and vegetables left on the table were ever taken after that. Even the baskets that held the food were left.

She didn't have a name for her creature; she would sometimes call him "the beast." After a while she began to call him King Kong, and that name stuck. It wasn't until the early 1970s that she found another name for her creature. There were a series of sightings and it was all over the news. "Bigfoot" he was called. Also known as "sasquatch" by the Indians. She couldn't pronounce "sasquatch," so she continued to call him King Kong, and only occasionally the bigfoot.

I had an encounter on my Bellingham property back in 1976. I never saw him, but I heard and smelled him and he made a mess of my garage. A half-hour after leaving my property, two security guards and two police officers chased a bigfoot into the woods at the small airport near my home. It was all over the news the following morning, so I knew for sure what was at my place that night. Gram said, "Well, you should've baked him a cake!"

Gram never mentioned a smell. She did occasionally mention hearing the howls and hoots in the summer time. I never heard or saw anything when I stayed at her place. It's almost like he knew she had company. Back in the year 2000, I did hear howls and shrieks myself while living at another country home near Granite Falls, Washington—eerie to say the least.

My wonderful, loving, caring, and sharing Gram passed away in November of 1989. Before she passed she gave me her "sasquatch cake" recipe, the very same one she made especially for HIM, and THEM, as she came to believe he was not always alone and may have had a family.

Now with you I share this special recipe, loved by her family, friends, and sasquatch himself. And if you should ever happen to have your own sasquatch pass through, please be sure to share with him.

Gram's Sasquatch Cake Recipe

Three eggs (beat two minutes)
One cup oil (for a richer taste use real butter, softened)
Two cups squash or pumpkin (your choice)
One teaspoon pumpkin spice
Two cups sugar
Three cups flour
One teaspoon salt
1/4 teaspoon baking powder
Three teaspoons baking soda
Three teaspoons vanilla
One cup nuts and one cup raisins or dried cranberries (I add all)

Mix well in your favorite large bowl. Lightly grease and flour a large bundt cake pan, or you can use several meatloaf pans (sasquatch prefer the bundt pan as they eat the cake like a donut). Bake at 325 degrees F. until done—depends on pan and oven—approximately one hour for a bundt pan, less for the meatloaf pans). Keep checking. Push on top until the cake springs back; use toothpick to check. Serve with coffee, milk or tea; enjoy with family and sasquatch friends.

The back of Gram's house and her garden where the sasquatch raids occurred.

Gram and her clothesline (above), and with a giant squash (right).

Blankets hung to dry and succulent vegetables appear to have been too much to resist for her unusual visitors.

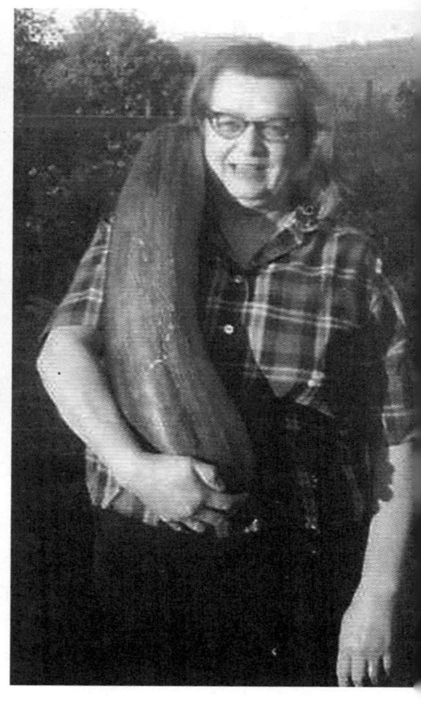

Gram with her "bigfoot-size" cabbages, and with a new harvest (inset). She was a "green-thumb" beyond belief. Everything she planted sprung from the ground in profusion. Her garden was like a gigantic cornucopia. It was no wonder that she was singled out for a free meal. I miss her very much.

Chapter 2

The Classics

The most beautiful thing we can experience is the mysterious.

—Albert Einstein

The stories presented below have been carried down through the generations; they come from long-ago media reports, official documents, or the case files of bigfoot researchers.

A warning to those who fear the darkness of the forests, the unexplained shriek that pierces the night, the snap of a twig, the rustling of the bushes, the shadow that passes by in the moonlight. A few of these stories scared even us. Yes us, believers of more kind and gentle bigfoot.

Before you venture further into this particular chapter, we ask that you take note and consider the period that most of these reports were documented. In these early days, horror associated with what was considered strange phenomenon was very popular. The media wishing to embellish and present fear rather than the soft or mushy side of a story was common.

As such, we accept these "classic stories" with our own set of "skepticals" on. That being said, the following tales are presented here for your education and analysis, as well as for their entertainment value. We must admit that most of the following tales only tickled our nerve endings, bringing a quiet sensation of nervous pleasure, while others left us with a deep sense of uneasiness about venturing into the woods on a dark night. But we do love an incredible tale, and even better, a scary story!

David Thompson, 1811

Although Indian sightings and legends of sasquatch go back hundreds of years, "documented" sightings in the Pacific Northwest date as far back as the early 18th century; 1811 seems to hold the earliest documented report. David Thompson, who has been called the greatest land geographer the world has ever known, wrote concerning his travels from 1792 to 1812 in the Western area of what is now the northern United States and southern Canada.

> From the numerous remains in Siberia and parts of Europe of the Elephant, Rhinocerous, and other large animals, especially near the Rivers, and in their banks, of those countries, I was led to expect to find the remains of those Animals in the Great Plaines and the Rivers that flow through them: but all my steady researches, and all my enquiry's led to nothing.

One thing that Thompson did find was an extraordinary set of footprints. The story appears under the entry for January 7, 1811, in his narrative. At the time, he was attempting to cross the Rocky Mountains via the valley of the Athabaska River and the Yellowhead Pass, and was in the vicinity of the present site of Jasper, Alberta—the first white men known to have followed what is now the route of the Canadian National Railway and the Yellowhead Highway.

> Continuing our journey in the afternoon we came on the track of a large animal, the snow about six inches deep on the ice: I measured it: four large toes each of four inches in length to each a short claw; the ball of the foot sunk three inches lower than the toes, the hinder part of the foot did not mark well, the length fourteen inches, by eight inches in breadth, walking from north to south, and having passed about six hours. We were in no humor to follow him: The Men and Indians would have it to be a young Mammoth and I held it to be the track of a large old grizzled bear; yet the shortness of the nails, the ball of the foot, and the great size was not that of a bear, otherwise that of a very large old Bear, his claws worn away; this the Indians would not allow.

Thompson referred to the incident again in his account of the return journey in the autumn:

> I now recur to what I have already noted in the early part of last winter, when proceeding up the Athabaska River to cross the Mountains, in company with…Men and four hunters, on one of the channels of the River we came to the track of a large animal, which measured fourteen inches in length by eight inches in breadth by a tape line. As the snow was about six inches in depth the track was well defined, and we could see it for a full one hundred yards from us, this animal was proceeding from north to south. We did not attempt to follow it, we had no time for it, and the Hunters, eager as they are to follow and shoot every animal made no attempt to follow this beast, for what could be the balls of our fowling guns do against such an animal. Report from old times had made the head branches of this River, and the Mountains in the vicinity the abode of one, or more, very large animals, to which I never appeared to give credence; for these reports appeared to arise for that fondness for the marvelous so common to mankind: but the sight of the track of that large beast staggered me, and I often thought of it, yet never could bring myself to believe such an animal existed, but thought it might be the track of some monster bear.

(Source: David Thompson, 1811. *David Thompson Narrative* [unpublished].)

There are no known images of David Thompson. The only "official" image is shown on this Canadian postage stamp, issued in 1957.

Note: Thompson's mention of four toes with claws, and failure to indicate that the creature was bipedal raises some doubt that the creature was a sasquatch. However, there have been other four-toed prints found that have been attributed to the creature, and claws indicated in at least one case.

The Diary of Elkanah Walker, 1840

This is an excerpt from the diary of Elkanah Walker, a pioneer and missionary in Washington State to the Spokane Indians. His established mission was approximately twenty-five miles northwest of present day Spokane, Washington. His mention of the snow-peaked mountain is believed to be Mt. Rainier:

> Bare *[sic]* with me if I trouble you with a little of their superstitions. They believe in a race of giants, which inhabit a certain mountain off to the West of us. This mountain is covered with perpetual snow. They [the creatures] inhabit the snow peaks. They hunt and do all their work at night. They are men stealers. They come to the peoples' lodges at night when the people are asleep, and take them and put them under their skins, and to take to their place of abode without even waking. Their track is a foot and a half long.
>
> We are told that the creatures steal salmon from Indian nets and eat them raw as the bears do. If the people are awake, they always know when they are coming very near by their strong smell that is most intolerable. It is not uncommon for them to come in the night and give three whistles and then the stones will begin to hit their houses.

The Story of Jacko, 1884

In 1884, an intriguing article appeared in *The Colonist,* a Victoria, British Columbia, newspaper. The article states that a creature, "something of the gorilla type," had been captured near Yale, British Columbia. The creature is described as standing about 4 feet, 7 inches (1.4 m) in height and weighing 127 pounds (57.5 kg). From these measurements, we might conclude that it was a young sasquatch. The following is an exact reprint of the article, with one exception; the actual newspaper article shows the date as 1882. This is an obvious error that I have corrected.

What Is It?
A Strange Creature Captured Above Yale

A British Columbia Gorilla

Correspondence to *The Colonist*

Yale, BC. July 3, 1884. In the immediate vicinity of No. 4 tunnel, situated some twenty miles above this village, are bluffs of rocks which have hitherto been unsurmountable, but on Monday morning last were successfully scaled by Mr. Onderdonk's employees on the regular train from Lytton. Assisted by Mr. Casterton, the British Columbia Express Company's messenger, and a number of gentlemen from Lytton and points east of that place who, after considerable trouble and perilous climbing, succeeded in capturing a creature which may truly be called half man and half beast.

"Jacko," as the creature has been called by his captors, is something of the gorilla type standing about four feet seven inches in height and weighing 127 pounds. He has long, black, strong hair and resembles a human being with one exception, his entire body, excepting his hands, (or paws) and feet are covered with glossy hair about one inch long. His fore arm is much longer than a man's fore arm, and he possesses extraordinary strength, as he will take hold of a stick and break it by wrenching or twisting it, which no man living could break in the same way. Since his capture he is very reticent, only occasionally uttering a noise which is half bark and half growl.

He is, however, becoming daily more attached to his keeper, Mr. George Tilbury, of this place, who proposes shortly starting for London, England, to exhibit him. His favorite food so far is berries, and he drinks fresh milk with evident relish. By advice of Dr. Hannington raw meats have been withheld from Jacko, as the doctor thinks it would have a tendency to make him savage.

The mode of capture was as follows: Ned Austin, the

engineer, on coming in sight of the bluff at the eastern end of the No. 4 tunnel saw what he supposed to be a man lying asleep in close proximity of the track, and as quickly as thought blew the signal to apply the breaks. The brakes were instantly applied, and in a few seconds the train was brought to a standstill. At this moment the supposed man sprang up, and uttering a sharp quick bark began to climb the steep bluff.

Conductor R.J. Craig and Express Messenger Custerton, followed by the baggageman and brakemen, jumped from the train and knowing they were some twenty minutes ahead of time gave immediate chase. After five minutes of perilous climbing the then supposed demented Indian was corralled on a projecting shelf of rock where he could neither ascend or descend. The query now was how to capture him alive, which was quickly decided by Mr. Craig, who crawled on his hands and knees until he was about forty feet above the creature. Taking a small piece of loose rock he let it fall and it had the desired effect on rendering poor Jacko incapable of resistance for a time at least. The bell rope was then brought up and Jacko was now lowered to terra firma. After firmly binding him and placing him in the baggage car "off brakes" was sounded and the train started for Yale.

At the station a large crowd who had heard of the capture by telephone from Spuzzum Flat were assembled, each one anxious to have the first look at the monstrosity, but they were disappointed, as Jacko had been taken off at the machine shops and placed in charge of his present keeper.

The question naturally arises, how came the creature where it was first seen by Mr. Austin? From bruises about its head and body, and apparent soreness since its capture, it is supposed that Jacko ventured too near to the edge of the bluff, slipped, fell and lay where found until the sound of the rushing train aroused him.

Mr. Thos. White and Mr. Gouin, C.E., as well as Mr. Major, who kept a small store about half a mile west of the tunnel during the past two years, have mentioned having seen a curious creature at different points between Camps 13 and 17, but no attention was paid to their remarks as people came to the conclusion that they had either seen a bear or stray Indian dog. Who can unravel the mystery that now surrounds Jacko? Does he belong to a species hitherto unknown in this part of the continent, or is he really what the train man first thought he was, a crazy Indian?

Note: Subsequent newspaper articles on this event indicate that the entire story was a hoax. However, we have information provided by a game guide, Chilco Choate, who stated his grandfather was there when this "ape" was brought in and kept at Yale. Whether grandfather Choate actually saw the creature is uncertain. Next we have a Mrs. Hillary Foskett, who stated that her mother was in Yale at the time (she was about 8 or 9 years old), and remembered stories of the creature. A Dr. Hannington, who is mentioned in The Colonist *article, was well known to her. In this connection, it has been established that all of the people mentioned in the article were real people. Lastly, Ellen Neal (who carved the totem poles in the Harrison Hot Springs Hotel lobby—Harrison is near Yale) was told by Chief August Jack Khahtsalano that a creature of this nature did reach Burrard Inlet in Vancouver and was exhibited there. John Green states that he was told that Chief Khahtsalano actually saw a creature (i.e., a sasquatch-like creature) on display in 1884. It is reasonable to assume that this was Jacko. Nevertheless, the story could have been fabricated—hoaxes were commonly practiced at that time.*

These illustrations provide some insights into the saga of Jacko. That the incident might not have actually occurred cannot be denied, but that the story is firmly entrenched in sasquatch lore is beyond doubt.

The Chetco River Incident, 1890

In 1890, a dozen loggers were cutting timber and camping with their families in an area near the Chetco River, in Southern Oregon. Each morning the loggers awoke to find enormously large, human-like footprints in the damp earth along the riverbank. Nearby shrubs and saplings were uprooted and whacked to shreds, and their freshly cut logs scattered and tossed about like matchsticks. Some logs were used to beat against machinery, which gave them serious doubts that the culprits might be bears.

One night, they were awakened by loud, inhuman shrieks and screams coming from the woods. From every tent, men bounded out of bed while grabbing their guns. No one lighted a lamp for fear of attracting the beast, and frightened children were warned not to cry. The spine-chilling noises went on and on. Sometimes they seemed close by, other times from the direction of the road or the river. Finally the sounds faded into the distance, and quiet returned to the dark campsite.

One of the men told this story to his fellow loggers; he had heard the story from a white man whom the Indians trusted enough to take into their confidence.

> Indians believed there were man-animals in the woods. The Chetco claimed that for generations they had shared their hunting grounds with fierce-looking hairy creatures who walked upright like men. The strange beings were not human, nor animal, neither friendly nor hostile. They were simply there, like every other man or wild creature, so the Indians left them alone.

Very late on the third night, the frightening sounds were once again heard faintly from far off in the woods. As the shrieks and screaming grew louder, every man in every tent pulled on their trousers and boots, and readied their guns. Obviously, the night screamer was coming closer and closer.

When the beast seemed only fifty feet away, one man took desperate action, hastily fashioning a torch of oily rags and kindling, and set fire to it. Torch in one hand and rifle in the other, he raced into the woods. In the meantime, the man's wife called for help. Within minutes, several men stumbled toward her in the darkness.

They groaned when they learned that their comrade had gone into the woods alone. None hesitated to follow, while one dashed off to fetch a lantern and others grabbed more ammunition. Finally, the party headed into the forest in the direction from which the awful sounds were heard.

Then it let out a tremendous roar...

They had covered only a short distance when the shrieking and screaming stopped. The men halted, and listened. There was a long silence, followed by an outburst of bestial howling which was then followed by lingering human screams. Thinking their friend was being attacked, the men fought through the undergrowth, the man with the lantern in the lead.

Moments later their comrade appeared, a look of sheer terror on his face as he collapsed into their arms. At first he was too terrified to speak. His companions fired their guns to drive off the howler, then waited patiently for the man to gasp out the details. In a panic, he rambled on about a hairy monster, eight feet tall with yellow eyes, fangs, huge hands, and shoulders that were two axe-handles wide across.

He said the torchlight must have blinded it because it stood stock-still, one hand shading its eyes. Then it let out a tremendous roar. The man hurled his torch into its face, but instead of shooting, he ran screaming back toward camp. The men did not doubt his word.

After returning to camp and much discussion, the loggers agreed to take turns standing guard day and night until the creature was captured, or shot. Firewood was gathered so that large fires could be kept blazing all night.

The next night, again hearing the shrieks and screams, two of the men decided they had had enough, and set out to track and kill the hairy creature. They headed off into the dark forest with their rifles and a lantern. Some time later, those at the campsite heard their shouts, screams, and gunshots. And then dead silence.

When the two men didn't appear, their friends grabbed their guns, lanterns, and torches, and firing shots into the air, they headed for the forest in search of their comrades. The searchers shouted, swung lanterns, and fired their guns so that their friends would know help was on the way. A half mile from camp they found their friends.

It was an incredibly gruesome sight that would live on in their nightmares forever. Blood dripped from the trees, shrubs, bushes, and even the branches high in the trees. The stench of blood filled the air. Body parts were scattered as the men's limbs had been torn from their bodies. They had been slammed hard against the trees and ripped apart, literally, limb from limb, by something incredibly strong. There was no sign of the creature, only bloody footprints leading away from the site and deeper into the forest.

The remaining loggers packed up and abandoned camp at first light. Professional hunters were sent back to the forest to search for the monster. But no sign of the creature (or creatures) was ever found.

Source: From the files of Ray Crowe.

Bigfoot Patrick, 1890

The following is a reprint of a story by Dr. Ed. Fusch in his book, *They Walked Among Us: Scweneyti and The Stick Indians of the Colvilles.*

> Most interesting, is the tale of Patrick. Before the turn of the century, the Lake Band of the Colville Indians, of Ferry County in Washington State, had set up a fishing camp near Keller on the San Poil River. A recent bride went for water, and a short time later was heard screaming. The men rushed to her aid but could only stand and watch as the "Skan-

icum" [sasquatch] carried the bride away.

She was with the Skanicum all summer, when searchers rescued her from the sleeping Skanicum as she gathered roots. During her stay, she became pregnant and bore a son named Patrick. His body was quite different; the five-foot four-inch tiny man had very long arms, reaching to his knees. He had a sloping forehead, a large lower jaw with a wide mouth, straight upper and lower lips, and straight protruding teeth.

He was hump-backed, his ears were peaked, he had long fingers and large hands, and generally when at school was described as very ugly, although very smart. He died at thirty, and is buried on the reservation. He did marry and was considered affluent. There were three daughters, and two sons which both died early.

The daughters were Mary Louise and Madeline (now in their seventies) and Stella (died young). Mary Louise lives near Omak, and has said that her paternal grandfather was a Skanicum. She is relatively normal in appearance, but both girls have wide mouths, protruding teeth, and squinted eyes.

Madeline, who lives on the Washington Coast, has a very distinct sloping forehead, and long peaked ears. She is considered ugly by Indian standards, and is an alcoholic, spending much time in taverns.

Note: Dr. Fusch is an anthropologist, geologist, and philosopher. His book is full of interesting tales told to him by friends in the Colville and Spokane Indian tribes.

The Bauman Story, 1892

The following is a reprint of a story by Theodore Roosevelt in his book, *The Wilderness Hunter.*

Frontiersmen are not, as a rule, apt to be very superstitious. They lead lives too hard and practical, and have too little imagination in things spiritual and supernatural. I have heard but few ghost stories while living on the frontier, and those few were of a perfectly commonplace and conventional. But I once listened to a goblin-story, which rather impressed me.

It was told by a grizzled, weather-beaten old mountain hunter, named Bauman, who was born and had passed all of his life on the frontier. He must have believed what he said, for he could hardly repress a shudder at certain points of the tale. But he was of German ancestry, and in childhood had doubtless been saturated with all kinds of ghost and goblin lore, so that many fearsome superstitions were latent in his mind; besides, he knew well the stories told by the Indian medicine-men in their winter camps, of the snow-walkers, and the specters, and the formless evil beings that haunt the forest depths, and dog and waylay the lonely wanderer, who after nightfall passes through the regions where they lurk; and it may be that when overcome by the horror of the fate that befell his friend, and when oppressed by the awful dread of the unknown, he grew to attribute, both at the time and still more in remembrance, weird and elfin traits to what was merely some abnormally wicked and cunning wild beasts; but whether this was so or not, no man can say.

When the event occurred, Bauman was still a young man, and was trapping with a partner among the mountains dividing the forks of the

Theodore Roosevelt

Theodore Roosevelt heard this story from Bauman while hunting in the Bitterroot Mountains. Bauman was by then an old man, so we believe the event occurred in the 1850s.

Salmon from the head of Wisdom River. Not having had much luck, he and his partner determined to go up into a particularly wild and lonely pass through which ran a small stream said to contain many beaver.

The pass had an evil reputation because the year before, a solitary hunter who had wandered into it was there slain, seemingly by a wild beast, the half eaten remains being afterwards found by some mining prospectors who had passed his camp only the night before.

The memory of this event, however, weighted very lightly with the two trappers, who were as adventurous and hardy as others of their kind. They took their two lean mountain ponies to the foot of the pass where they left them in an open beaver meadow, the rocky timber-clad ground being from there onward impracticable for horses.

They then struck out on foot through the vast, gloomy forest, and in about four hours, reached a little open glade where they concluded to camp, as signs of game were plenty. There was still an hour to two of daylight left, and after building a brush lean-to and throwing down and opening their packs, they started upstream.

The country was very dense and hard to travel through, as there was much down timber, although here and there the somber woodland was broken by small glades of mountain grass. At dusk they again reached camp. The glade in which it was pitched was not many yards wide, the tall, close-set pines and firs rising round it like a wall. On one side was a little stream, beyond which rose the steep mountain slope, covered with the unbroken growth of evergreen forest.

They were surprised to find that during their absence, something, apparently a bear, had visited camp, and had rummaged about among their things, scattering the contents of their packs, and in sheer wantonness, destroying their lean-to. The footprints of the beast were quite plain, but at first they paid no particular heed to them, busying themselves with rebuilding the lean-to, laying out their beds and stores and lighting the fire.

While Bauman was making ready supper, it being

already dark, his companion began to examine the tracks more closely, and soon took a brand from the fire to follow them up, where the intruder had walked along a game trail after leaving the camp. When the brand flickered out, he returned and took another, repeating his inspection of the footprints very closely.

Coming back to the fire, he stood by it a minute or two, peering out into the darkness, and suddenly remarked, “Bauman, that bear has been walking on two legs.” Bauman laughed at this, but his partner insisted that he was right, and upon again examining the tracks with a torch, they certainly did seem to be made by but two paws or feet.

However, it was too dark to make sure. After discussing whether the footprints could possibly be those of a human being, and coming to the conclusion that they could not be, the two men rolled up in their blankets, and went to sleep under the lean-to. At midnight Bauman was awakened by some noise, and sat up in his blankets.

As he did so his nostrils were struck by a strong, wild-beast odor, and he caught the loom of a great body in the darkness at the mouth of the lean-to. Grasping his rifle, he fired at the vague, threatening shadow, but must have missed, for immediately afterwards he heard the smashing of the under wood as the thing, whatever it was, rushed off into the impenetrable blackness of the forest and the night. After this the two men slept but little, sitting up by the rekindled fire, but they heard nothing more.

In the morning, they started out to look at the few traps they had set the previous evening and put out new ones. By an unspoken agreement, they kept together all day, and returned to camp toward evening.

On nearing it they saw, hardly to their astonishment, that the lean-to had again been torn down.

The visitor of the preceding day had returned, and in wanton malice had tossed about their camp kit and bedding, and destroyed the shanty. The ground was marked up by its tracks, and on leaving the camp it had gone along the soft earth by the brook, where the footprints were as plain as if on snow, and, after a careful scrutiny of the trail, it certainly did seem as if, whatever the thing was, it had walked on two legs. The men, thoroughly uneasy, gathered a great heap of dead logs and kept up a roaring fire throughout the night, one or the other sitting on guard most of the time.

About midnight the thing came down through the forest opposite, across the brook, and stayed there on the hillside for nearly an hour. They could hear the branches crackle as it moved about, and several times it uttered a harsh, grating, long-drawn moan, a peculiarly sinister sound. Yet, it did not venture near the fire. In the morning the two trappers, after discussing the strange events of the last thirty-six hours, decided that they would shoulder their packs and leave the valley that afternoon.

They were the more ready to do this because in spite of seeing a good deal of game sign, they had caught very little fur. However, it was necessary first to go along the line of their traps and gather them, and this they started out to do. All the morning they kept together, picking up trap after trap, each one empty.

On first leaving camp, they had the disagreeable sensation of being followed. In the dense spruce thickets they occasionally heard a branch snap after they had passed; and now and then there were slight rustling noises among the small pines to one side of them. At noon, they were back within a couple miles of camp.

In the high, bright sunlight, their fears seemed absurd to the two armed men, accustomed as they were, through long years of lonely wandering in the wilderness, to face every kind of danger from man, brute or element. There were still three beaver traps to collect from a little pond in a wide ravine near by.

Bauman volunteered to gather these and bring them in, while his companion went ahead to camp and made ready the packs. On reaching the pond, Bauman found three beavers in the traps, one of which had been pulled loose and carried into a beaver house. He took several hours in securing and preparing the beaver, and when he started homewards he marked, with some uneasiness, how low the sun was getting.

As he hurried toward camp, under the tall trees, the silence and desolation of the forest weighted on him. His feet made no sound on the pine needles and the slanting sun-rays, striking through among the straight trunks, made a gray twilight in which objects at a distance glimmered indistinctly.

There was nothing to break the gloomy stillness which, when there is no breeze, always broods over these somber primeval forests. At last he came to the edge of the little glade where the camp lay, and shouted as he approached it, but got no answer. The campfire had gone out, though the thick blue smoke was still curling upwards.

Near it lay the packs wrapped and arranged. At first Bauman could see nobody; nor did he receive an answer to his call. Stepping forward he again shouted, and as he did so his eye fell on the body of his friend, stretched beside the trunk of a great fallen spruce. Rushing toward it, the horrified trapper found that the body was still warm, but that the neck was broken, while there were four great fang marks in the throat.

The footprints of the unknown beast-creature, printed deep in the soft soil, told the whole story. The unfortunate man, having finished his packing, had sat down on the spruce log with his face to the fire, and his back to the dense woods, to wait for his companion.

While thus waiting, his monstrous assailant, which must have been lurking in the woods, waiting for a chance to catch one of the adventurers unprepared, came silently up from behind, walking with long noiseless steps and seemingly still on two legs. Evidently unheard, it reached the man, and broke his neck by wrenching his head back with its fore paws, while it buried its teeth in his throat.

It had not eaten the body, but apparently had romped and gamboled around it in uncouth, ferocious glee, occasionally rolling over and over it; and had then fled back into the soundless depths of the woods.

Bauman, utterly unnerved, and believing that the creature with which he had to deal was something either half human or half devil, some great goblin-beast, abandoned everything but his rifle and struck off at speed down the pass, not halting until he reached the beaver meadows where the hobbled ponies were still grazing. Mounting, he rode onwards through the night, until beyond reach of pursuit.

Note: Theodore Roosevelt later became president of the United States. The Bauman story is considered the first major published report of a possible bigfoot encounter.

The Albert Ostman Story, 1924

In 1924, a Canadian lumberjack by the name of Albert Ostman was prospecting near Toba Inlet when he says he was kidnapped by a bigfoot and held captive by the creature's family. In the middle of the night he was picked up, sleeping bag and all, and carried off a good three miles to the home and camp site of a family of sasquatch, which included the father who had brought him there, the mother and two children—a boy and a girl. The creatures made no effort to hurt him, but it was clear they did not want him to escape.

A drawing of the creature Ostman encountered, prepared under his direction.

Ostman's description went like this:

> The old Man was eight feet tall with a big barrel chest and big hump on his back—powerful shoulders, the biceps on his upper arms were enormous and tapered down to his elbows. His forearms were longer than common people have, but well proportioned. His hands were wide, the palm was long and broad and hollow like a scoop. His fingers were short in proportion to the rest of the hand.

Ostman described the son as almost full-grown, seven feet tall and about three hundred pounds. The daughter was young and always kept her distance from Ostman. All were covered in hair except for the palms of their hands, soles of their feet, and an area at the upper part of the nose and around the eyelids.

Living with them for nearly a week, Ostman learned much about the creatures. They had a language all their own and would chatter among themselves. They seemed to be vegetarians, as he never saw them eat meat nor do any cooking. They gathered some kind of nut that grows in the ground, and roots that have a very sweet and satisfying taste. He thought them to move from place to place, as food is available in different localities. They slept under homemade blankets made of cedar bark strips woven together.

He was never harmed in any way, but after six days, Ostman was planning an escape. When the old man grabbed his snuff can and swallowed the whole contents, thereby becoming very sick, and then grabbed and downed Ostman's pot of hot coffee, grounds and all, Ostman was able to slip away. Fearing people would think him crazy, it was thirty-four years later that he finally told his story which was sworn before a justice of the peace in Fort Langley, British Columbia, on August 20, 1957.

Albert Ostman is seen in the above photograph being interviewed by John Green in the 1950s. Below is Ostman's sworn statement that the event he relat ed was true. He is no longer wit us, but he never relented in the least as to his experience, and it remains one of the main accoun in sasquatch/bigfoot history.

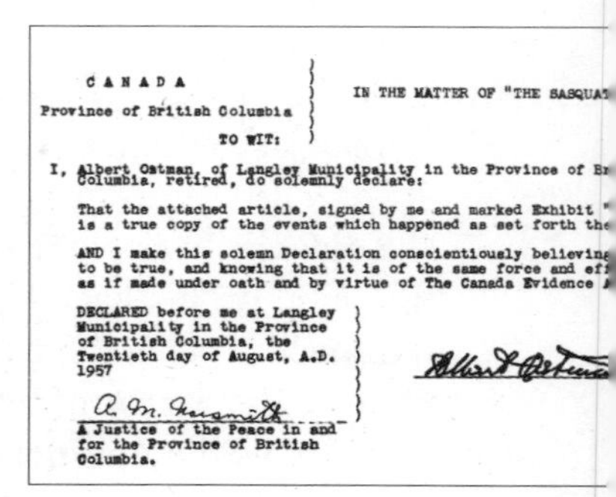

CANADA
Province of British Columbia
TO WIT:

IN THE MATTER OF "THE SASQUAT

I, Albert Ostman, of Langley Municipality in the Province of Br
Columbia, retired, do solemnly declare:

That the attached article, signed by me and marked Exhibit "
is a true copy of the events which happened as set forth the

AND I make this solemn Declaration conscientiously believing
to be true, and knowing that it is of the same force and ef
as if made under oath and by virtue of The Canada Evidence A

DECLARED before me at Langley Municipality in the Province of British Columbia, the Twentieth day of August, A.D. 1957

A. M. Naismith
A Justice of the Peace in and for the Province of British Columbia.

Albert Ostman

(Source: John Green, 2006. *Sasquatch: The Apes Among Us* [Second Edition], pp. 99–110.)

The Ape Canyon Incident, 1924

The first most publicized encounter in Washington State took place in the summer of 1924 when Fred Beck and four other prospectors state they were attacked by a number of sasquatch.

The men had been prospecting in the Mt. St. Helens and Lewis River area (southern Washington State) for about six years. They had staked a gold claim, which they named the Vander White, about two miles east of Mt. St. Helens. Here, they built a cabin near a deep canyon. Occasionally, they saw large footprints, which as far as they knew did not match those of any known animals. The largest print they observed measured 19 inches.

One evening they heard peculiar whistling and thumping sounds that continued for about one week. Later, while Beck and one of the other men were getting water at a nearby spring, the two observed a strange creature about 100 yards away. The other man took three rifle shots at the creature, which quickly disappeared. When it reappeared, about 200 yards away, Beck also took three shots before the creature again disappeared.

After the other men were informed of this incident, all agreed to go home the following morning. That night, however, several of the creatures attacked the cabin. They started by pelting it with rocks. As there were no windows in the cabin, the men could not see the assailants. The men's only view outside was through a chinking space. With the limited field of view and the darkness, nothing was actually seen. (In his book, Beck states that three creatures were seen, however, this was later found to be incorrect.)

Later the creatures climbed on the cabin roof and tried to break down the door. The men fired their rifles through the roof and through the door. One creature even reached into the cabin through the chinking space and grabbed hold of an ax. Marion Smith, Beck's father-in-law, turned the ax head so that it caught on the logs. He then shot his rifle along the ax handle, and the creature let go of the ax.

The attack ended just before daylight. When it was light enough, the men ventured outside. A short time later, Beck saw one of the creatures about eighty yards away near the edge of the canyon. He took three shots at it and saw it topple into the gorge, which was about 400 feet deep. The men then hastily left the area without pack-

Fred Beck is seen in the first photograph much later in life with his trusty rifle. The group seen top right shows investigators standing by the besieged cabin. From left to right: Burt Hammerstrom, freelance writer; Bill Welch, forest guard at the Spirit Lake Ranger Station; Frank (Slim) Lynch, Seattle newsman; and Jim Huffman, forest ranger for the Spirit Lake district. Lower left shows Beck (left) and Roy Smith, one of the other miners. They are at the cabin re-enacting the scene for the investigators. On the lower right is Ape Canyon, so named for this unusual event. We are told a set of large footprints was found, but, oddly, no photograph was taken.

ing their supplies and equipment. They took only what they could carry in their packsacks.

There are claims that this entire event was a hoax played on the miners. However, there is no explanation for Beck's claim that he shot one of the creatures. Beck later stated that the creatures were paranormal entities from another dimension.

(Source: Fred Beck, 1967. *I Fought the Apemen of Mt. St. Helens,* written by Ronald Beck, This summary is from *Meet the Sasquatch* by Christopher L. Murphy with John Green and Thomas Steenburg.)

The Muchalat Harry Story, 1928

The following is a reprint of the account in Peter Byrne's book, *The Search for Big Foot: Monster, Myth or Man?* pp. 1–6.

According to the Indians, there were once a large number of Bigfeet living on Vancouver Island, a large island, 12,408 square miles in area, off the West Coast of British Columbia. The Indians knew about them, feared them, and respected them, but granted that they were harmless. One of the Indians of the Nootka Tribe, who lived at Nootka in 1928, claims to have been carried off by them and held captive for some time.

Peter Byrne

The story, told to me by Father Anthony Terhaar of Mt. Angel Abbey in Oregon, is a curious one. Father Anthony, a much-loved missionary priest who traveled the West Coast of Vancouver Island for many years, was living at Nootka at the time of the story and he knew Muchalat Harry very well.

Muchalat Harry was a trapper and something of a rarity among his fellow tribesmen. He was, according to Father Anthony, a tough, fearless man, of excellent physique. In the course of his trapping, he was wont to spend long weeks in the forest alone, something that the average Indian did not do in those days. The Indians of the coast were apparently a rather timid people and they seemed to regard the deep forests as the home and territory of the Bigfoot. When they went into the deep inland forest for any reason, they never went alone. Muchalat Harry was different from other Indians. He went in the forests alone and feared nothing.

The Conuma River where it enters Tiupana Inlet on the west coast of Vancouver Island. It was here where Muchalat Harry cached his canoe for the last time in 1928.

Late one autumn Muchalat Harry set off for the woods with his traps and camping gear. His plan was to set out a trap line and stay in the woods for several months. He headed for his favorite hunting area, the Conuma River, at the head of Tlupana Inlet. From Nootka he paddled his own canoe to the mouth of the Conuma. There he cached the canoe and headed upstream on foot. Approximately twelve miles upstream he made his base camp and, after building himself a lean-to, started to put out his trap line.

One night, while wrapped in his blankets and clad only in his underwear, he was suddenly picked up by a huge male Bigfoot and carried off into the hills. He was not carried very far, probably a distance of two or three miles, at the most. When daylight came, he was able to see that he was in a sort of camp, under a high rock shelf and surrounded by some twenty Bigfeet. They were of all sexes and sizes. For some time they stood around him and stared at him. The males were to the front of the curious group, females behind them, and young ones to the rear. Muchalat Harry was frightened at first and his fear grew to terror when he noticed, he said, the large number of bones lying around the campsite. When he saw these he was convinced that the Bigfeet were going to eat him.

The Bigfeet did not harm him in any way. Occasionally one came forward and touched him, as if feeling him, and when they discovered that his "skin" was loose—it was in fact his woolen underwear—several came forward and pulled at it gently.

While they looked at him and examined him, Muchalat Harry sat with his back to the rock wall and did not move. He was cold and hungry, but his thoughts were only on escape. Sometime in the late afternoon, curiosity on the part of the Bigfeet seemed to slacken and with most of the Bigfeet out of camp, probably food-gathering he thought, there came the opportunity that he needed. He leapt to his feet and ran for his life, never looking back. He ran downhill, toward where he guessed the river to be and sure enough, he soon came to his campsite. In what must have been blind panic he bypassed his camp and ran for twelve miles to where his canoe was cached at the mouth of the Conuma.

Map of Vancouver Island showing Muchalat Inlet (circled) where Muchalat Harry was abducted.

The story of Muchalat Harry's arrival at Nootka is described by Father Anthony as follows:

It was probably three in the morning. He and his brother Benedictines were asleep and the village was quiet. Suddenly there was a series of wild cries from the waters of the inlet. Lights were lit and he and others hurried down to the water's edge. There, near-frozen and exhausted in his canoe, lay Muchalat Harry. He was barefoot and clad only in his wet and torn underwear, and he had paddled his canoe through the winter night forty-five miles from the mouth of the Conuma River.

Father Anthony Terhaar

Father Anthony and his companions carried the almost lifeless form up from the water's edge. It took three weeks to nurse Muchalat Harry back to sanity and good health. The nursing was done by Father Anthony, who took him into his own care, and he told me that during the course of these three weeks, Muchalat Harry's hair turned to pure white.

The story of the kidnapping came out slowly. At first Muchalat Harry would talk to no one. Then he told Father Anthony what had happened and, later, others. When he was fully recovered to health he was asked when he planned to go back to collect his belongings, the camp equipment, the pots and pans, his trap line and above all, his rifle, at the lean-to on the Conuma. In 1928, a trap line and all of its pieces must have been worth a great deal to an island Indian. A rifle alone would be regarded as a highly prized possession. But Muchalat Harry never went back to the Conuma. Not only did he never return there; according to Father Anthony, he never left the settlement at Nootka, never went in the woods again for the rest of his life. He preferred to lose all of his valuable and probably hard won possessions rather than risk another encounter with the Bigfeet.

Note: Peter Byrne has been involved in the search for bigfoot since the late 1950s. He headed several major initiatives to find firm evidence supporting the creature's existence. He is still active in bigfoot research. He recently moved to Oregon to continue what he terms, The Great Search.

The Coppei Falls Bigfoot, 1920s

The late Vance Orchard wrote stories about people, places, and bigfoot in southeast Washington and northeast Oregon for many years. One of the most interesting by far was from a lady in her eighties, who wishes to remain anonymous.

She told of her childhood days in the 1920s, growing up on the upper Coppei, an area in the foothills of the Blue Mountains near Walla Walla, Washington in Columbia County. There at Coppei Falls, she said bigfoot would chase elk and run them over the brink at the falls. Then, at the base of the cliff, they would pack off the elk at their leisure. Early Native Americans had used the same tactic to kill thousands of buffalo.

She had a number of stories to tell about a whole family of the creatures. Members of the family often raided her family's garden, as well as the gardens of other locals. She remembered following tracks left by the barefooted creatures. Sometimes the stride of the tracks were so far apart she had to run and jump to get from one step to another. "One night, a real big one looked into our windows," she said. Lots of the locals had the same experience. Having the creatures around was pretty unnerving to most of the people who lived in the area, and many of them sold off and moved away.

One day, while walking about the property she ran into one of the man-creatures, this one about six feet tall. It was reaching up into a young tree, its fingers spread. It had brown hair and its face was gray. It turned its head on its shoulders like an owl would to look at her. She froze where she stood until it walked away, disappearing into the brush.

Her mother had an encounter one day when she found the family herd of six milk cows missing from the pasture. She noticed their hoof tracks and followed them, after a while catching up to them. They were not alone; with them was a young bigfoot. She looked at the bigfoot and it looked at her. She was scared, but was not about to leave her cows; she ran after the cows herding them back toward the homestead. Her mother later commented that, "The creature was just a caveman, but he was dressed up in furs and really needed a bath!"

One day, when she was thirteen years old, her father told her he was looking across an open hillside when he saw two of the biggest, strangest bears ever step out of the brush. Two others soon followed as they walked over to a huge log that rested on the ground. They sat on the log and proceeded tearing it apart for grubs and such. When they had their fill they simply walked off down the trail and disappeared into the woods.

She said, "The creatures were a family, no mistake about that, and I think they even had names for each other. And, another thing, when a bigfoot got angry, they'd throw rocks at whatever made them angry."

Source: Vance Orchard, 2001. *The Walla Walla Bigfoot*, p. 6.

The Chapman Incident, 1941

George Chapman, a native Canadian, lived with his wife, Jeannie, and their three children (eight, seven, and five years old) in a small isolated house on the banks of the Fraser River near Ruby Creek, British Columbia. George was employed as a railroad maintenance worker at Ruby Creek. In September of 1941, he was surprised to see his wife and children running down the railroad tracks toward him. Jeannie excitedly told her husband that a sasquatch was after her.

The Chapman's house near Ruby Creek many years after it had been deserted. The twin evergreen trees are still there, but the house eventually crumbled and was removed when a relative of the Chapmans, Deborah Schneider and her husband, David, built a new house on the property. It was built to the left (facing) of the twin trees.

It all started while one of the children was playing in the front yard of the house. The child came running into the house shouting that a "big cow" was coming out of the woods. Jeannie looked out and saw an ape-like creature, over seven feet tall, covered in dark hair, approaching the house. Terrified, she grabbed her children and fled.

George and other men went to the house and found sixteen-inch-long footprints that led to a shed where a heavy barrel of fish had been dumped out. The prints then led across a field and into the mountains. Footprints on each side of a wire fence, four to five feet high, gave another clue as to the size of the creature—it apparently just took the fence in stride.

René Dahinden is seen here climbing over the wire fence which the creature apparently "took in stride."

The Chapmans returned to their home, but were continually bothered by unusual howling noises and their agitated dogs (that appeared to sense an unusual presence). The family left the house within one week and never returned.

The incident was thoroughly investigated by researchers, and a cast was made of one of the creature's footprints (cast length was seventeen inches—casts are always larger than actual prints).

(Source: Christopher L. Murphy in association with John Green and Thomas Steenburg, 2004. *Meet the Sasquatch*, p.35.)

William Roe and the Mica Mountain Bigfoot, 1955

William Roe's encounter with a probable bigfoot is by far the most credible account on record (notwithstanding the Patterson/Gimlin film). What I present here is a reprint of his signed statement on the event he experienced (Exhibit A), followed by an image of his sworn affidavit on the the truthfulness of his account.

EXHIBIT A

Ever since I was a small boy back in the forest of Michigan, I have studied the lives and habits of wild animals. Later, when I supported my family in northern Alberta by hunting and trapping, I spent many hours just observing the wild

The long gentle slope of Mica Mountain is seen here in the immediate background. The photo was taken in about 2004 at the only (and very old) motel in Tete Jaune Cache. Very little changes in the region, and I would say the scene would have been about the same in 1955.

things. They fascinated me. But the most incredible experience I ever had with a wild creature occurred near a little town called Tete Jaune Cache, British Columbia, about eighty miles West of Jasper, Alberta.

I had been working on the highway near Tete Jaune Cache for about two years. In October 1955, I decided to climb five miles up Mica Mountain to an old deserted mine, just for something to do. I came in sight of the mine about three o'clock in the afternoon after an easy climb. I had just come out of a patch of low brush into a clearing, when I saw what I thought was a grizzly bear, in the bush on the other side. I had shot a grizzly near that spot the year before.

This one was only about seventy-five yards away, but I didn't want to shoot it, for I had no way of getting it out. So I sat down on a small rock and watched, my rifle in my hands. I could see part of the animal's head and the top of one shoulder. A moment later, it raised up and stepped out into the opening. Then I saw it was not a bear.

This, to the best of my recollection, is what the creature looked like and how it acted as it came across the clearing directly toward me. My first impression was of a huge man, about six feet tall, almost three feet wide, and probably weighing somewhere near three hundred pounds. It was covered from head to foot with dark brown, silver-tipped hair. But as it came closer, I saw by its breasts that it was female.

And yet, its torso was not curved like a female's. Its broad frame was straight from shoulder to hip. Its arms were much thicker than a man's arms, and longer, reaching almost to its knees. Its feet were broader proportionately than a man's, about five inches wide at the front and tapering to much thinner heels. When it walked, it placed the heel of its foot down first, and I could see the grey-brown skin or hide on the soles of its feet.

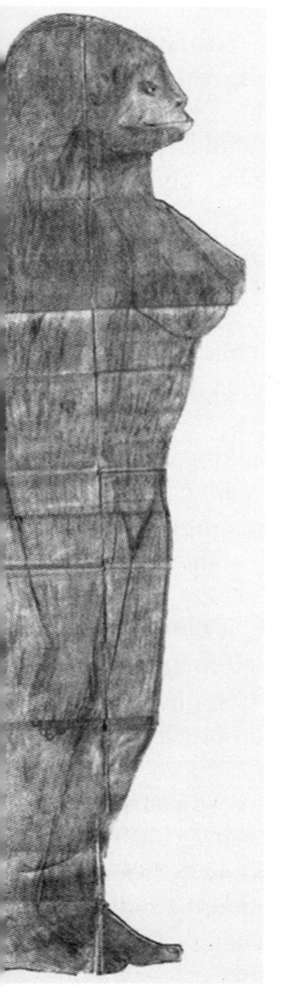

It came to the edge of the bush I was hiding in, within twenty feet of me, and squatted down on its haunches. Reaching out its hands, it pulled the branches of bushes toward it and stripped the leaves with its teeth. Its lips curled flexibly around the leaves as it ate. I was close enough to see that its teeth were white and even.

The shape of this creature's head somewhat resembled a Negro's. The head was higher at the back than at the front. The nose was broad and flat. The lips and chin protruded farther than its nose. But the hair that covered it, leaving bare only the parts of its face around the mouth, nose and ears, made it resemble an animal as much as a human. None of this hair, even on the back of its head, was longer than an inch, and that on its face was much shorter. Its ears were shaped like a human's ears. But its eyes were small and black like a bear's. And its neck also was inhuman. Thicker and shorter than any man's I had ever seen.

Drawing by Roe's daughter, Myrtle, made under his direction. There are some discrepancies with Roe's description, however, the drawing appears to be essentially accurate.

As I watched this creature, I wondered if some Movie Company was making a film at this place and that what I saw was an actor, made up to look partly human and partly animal. But as I observed it more, I decided it would be impossible to fake such a specimen. Anyway, I learned later there was no such company near that area. Nor, in fact, did anyone live up Mica Mountain, according to the people who lived in Tete Jaune Cache.

Finally, the wild thing must have got my scent, for it looked directly at me through an opening in the brush. A look of amazement crossed its face. It looked so comical at the moment I had to grin. Still in a crouched position, it backed up three or four short steps, then straightened up to its full height and started to walk rapidly back the way it had come. For a moment, it watched me over its shoulder as it went, not exactly afraid, but as though it wanted no contact with anything strange.

The thought came to me that if I shot it, I would possibly have a specimen of great interest to scientists the world over. I had heard stories of the Sasquatch, the giant hairy Indians that live in the legends of British Columbia Indians, and also many claim, are still in fact alive today. Maybe this was a Sasquatch, I told myself.

I leveled my rifle. The creature was still walking rapidly away, again turning its head to look in my direction. I lowered the rifle. Although I have called the creature "it," I felt now that it was a human being and I knew I would never forgive myself if I killed it.

Just as it came to the other patch of brush, it threw its head back and made a peculiar noise that seemed to be half laugh and half language, and which I can only describe as a kind of a whinny. Then it walked from the small brush into a stand of lodge pole pine.

I stepped out into the opening and looked across a small ridge just beyond the pine to see if I could see it again. It came out on the ridge a couple of hundred yards away from me, tipped its head back again, and again emitted the only sound I had heard it make, but what this half-laugh, half-language was meant to convey, I do not know. It disappeared then, and I never saw it again.

I wanted to find out if it lived on vegetation entirely or ate meat as well, so I went down and looked for signs. I found it in five different places, and although I examined it thoroughly, could find no hair or shells of bugs or insects. So I believe it was strictly a vegetarian.

I found one place where it had slept for a couple of nights under a tree. Now, the nights were cool up the mountain, at this time of year especially, and yet it had not used a fire. I found no sign that it possessed even the simplest of tools. Nor a single companion while in this place.

Whether this was a Sasquatch, I do not know. It will always remain a mystery to me, unless another one is found.

I hereby declare the above statement to be in every part true, to the best of my powers of observation and recollection.

(Signed) William Roe

AFFIDAVIT

I, W. Roe, of the City of Edmonton, in the Province of Alberta, MAKE OATH AND SAY:

1. That the Exhibit "A" attached to this my Affidavit is absolutely true and correct in all details .

SWORN before me in the City of Edmonton in the Province of Alberta this 26 day of August A.D. 1957) William Roe

William H. Clark
FU 2822

A Commissioner for Oaths in and for the Province of Alberta.

The Cowman of Copalis Beach, 1967

The following is a an account written by S.D. Baker of a story told to him by his father. The incident involves personal friends of the Bakers when they lived at Copalis Beach. S.D. Baker first relates the circumstances of the story and then provides a full narrative on the unusual events.

My dad worked in the timber industry his whole life. His father was a logger, and he grew up in and around the woods. My dad started his own logging company when he was eighteen, and has owned and operated shake and shingle mills from Oregon clear up to Thorn Bay, Alaska. He is an intelligent man and holds over a dozen patents for various pieces of equipment he has designed and built over the years. He has employed dozens of people over the years, all of them spending extensive time in the wilderness.

When I was a boy, I remember hearing bits and pieces of conversations among some of the men at the mill. Although nobody would tell me directly, I understood that something had gone on before I was born. They weren't joking around; they were genuinely afraid, and wouldn't talk about it with a kid. When I was young, my dad wouldn't tell me about it because I would often go out into the woods cutting blocks with him on the weekends, and he didn't want me to be afraid or develop a fear of the woods.

It was the first weekend of June 2004, that I was speaking with my dad about a couple of strange events that happened to me while out in the wilderness, and that reminded me of the hints at the story I heard when I was a boy. After some prodding he told me the following story:

In the mid 1960s, my dad owned a large roofing product mill in Aberdeen, Washington. He had teams of men who would cut the fallen old growth cedar salvage left after a logging operation. He had permits to salvage a large amount of wood in the coastal areas of Grays Harbor County, primarily in the area around Copalis Beach. Several of the men on his cutting crews lived in and around Copalis Beach. His foreman, a man I will call "Jon" for the story, was a bright, down to earth, hard worker. My dad trusted him with thousands of dollars of vehicles and equipment, as well as the safety of his crews. He was not the kind of man to make up stories.

On a Monday morning sometime in July, Jon was several hours late for work. This was highly unusual as he was always there early, getting the saws and trucks ready for the day. My dad said he was visibly shaken up when he arrived, and when he asked him what was wrong, Jon asked my dad to go in the office so the others wouldn't hear them. They went in and sat down, and Jon simply said, "Something destroyed our house this weekend."

My dad thought he said "someone" broke into the house, and asked Jon if it was someone he knew. Jon said, "You don't understand, this wasn't a person. It was a—I don't know what it was, but it completely trashed the house. The family is going to stay with my brother in Elma for a while."

My dad asked him to explain what had happened. Jon said that when he got home from work Friday evening, his youngest son, Tim, who was around four at the time, told him he saw a big "cowman" walking at the edge of their field that afternoon. He thought the boy meant "cowboy," because some of his neighbors wore cowboy hats when they were out in the sun.

He asked him if the man was wearing a cowboy hat, and the boy said, "No daddy, he was a Cowman, furry and stinky like the cows." He asked his wife if she knew what the boy was talking about. She said Tim was playing on the porch that afternoon, when he came running in and said the cowman was stuck on the fence. He was very excited, so she went out to see what he was talking about.

She said as she opened the door, she was hit by a horrible smell, like that of wet dog and garbage. Tim was pointing across to the field opposite their house and said, "He got loose!" She looked where he was gesturing and could see the top strand of barb wire bouncing up and down, as if somebody had just pulled on it really hard and let it go. She didn't see the "cowman," and noticed nothing out of the ordinary except for the smell.

Feeling uneasy and scared, she told Tim to come inside to play for rest of the day. Their older son, Jon Jr., who was twelve at the time, was at a friend's house and walked home a short while after Tim saw his "cowman." He told his mother somebody had followed him home, walking in the woods off to the right side of the road. He never saw who it was; [whatever it was] never left the woods, but he said it had to be a really big man.

He would hear large twigs cracking, and the footsteps were very heavy. Once he got to the driveway of their house where the woods stopped at the field where his brother had his sighting, the footsteps stopped and Jon Jr. never saw anything. He was pretty shaken up by the event, and wanted his dad to go out to the woods and check it out with him.

Later that evening, Jon strapped on his .357 and took his older son out into the field to have a look. They first walked to the area where the "cowman" was supposedly stuck on the fence, and walked down the fence line looking for anything. They came upon a large clump of long, reddish-brown hair tangled in the top strand of barbed wire. He tried to pull it off, but it was really tangled up, so he pulled out his buck knife and sawed it off. He said the hair was over a foot long, real coarse and stringy; the hair kind of looked like it was from a horse's mane or tail.

There appeared to be a bit of flesh matted in the clump, and the top wire was pulled loose from one of the posts. Whatever was hung up on the fence was very big. He handed the hair to his son to hold, and they climbed through the fence and walked toward the woods. He said he was looking for any sign of tracks on the ground. The ground was a solid grassy field, and there were no hoof prints or any other tracks he could see.

The edge of the woods began about ten feet from the fence line, and they entered on a small game trail that deer frequented. It was around eight at night, and in the woods it was getting to be fairly dark. They walked for a ways, and soon began to smell the rotting garbage/wet dog odor his wife reported earlier. Jon said he got the feeling they were being watched; the hair on the back of his neck was standing up. He told his son they should head back before it got dark, and the boy didn't argue.

As they began walking back out, they could hear heavy footsteps off to their left. They stopped, and the footsteps stopped. They walked on nearly to the clearing, and Jon whispered to his son to run like hell to the house on the count of three. Jon Jr. nodded, and Jon whispered, "One, two—three!" and gave his son a push in the back to get him started, then spun around and raced off the trail in the opposite direction, toward the footsteps, with his gun drawn.

Off the trail, the underbrush was dense with ferns and bushes; he had a hard time making headway. But as he got closer, he could hear it moving away from him, deeper into the woods. At this time, he told my dad that he thought it was a vagrant camping out in the woods and possibly scoping out houses to rob at night. Jon was a big man and capable of taking care of himself in most any situation, and he had a large caliber handgun, so he wasn't too worried about confronting a vagrant in the woods.

He was a few yards off the trail in deep brush when he heard the movement stop just ahead of him. He stopped to look and listen, and thought he saw movement by a large tree, like someone was trying to hide there. He leveled his gun and said "Come out nice and slow, or I swear to God I'll come back there and shoot you!"

It was silent for a moment, and then he caught movement out of the corner of his eye and spun around to his right for a better look. He said it looked like a huge bear moving through the brush; he could only see bits of it through the dense ferns, but it was moving quietly away from the tree on four legs. It was about fifteen feet away from him. At first he thought it was a bear, and then suddenly he saw a huge hairy arm with a human-like hand reach out of the brush and grab a small alder tree. The tree was about four inches in diameter, and it grabbed hold about five feet up.

He said it happened so fast it was a blur, but the thing pulled itself upright out of the brush by holding the tree. It stood on two legs and turned its upper body to glare at Jon. It was enormous; he couldn't believe how bulky it was. He said it was well over seven feet tall, and at least half that big through the chest. It was too dark to make out many features, but its eyes seemed to glow a deep red, and he thought he could see teeth, like it was curling its lips back.

It stood for just a brief moment, and then lunged ahead, pushing back on the tree with tremendous force. The tree snapped loudly and crashed into the trees around it, getting hung up in the branches and not falling to the ground. It then disappeared into the deep brush with frightening speed, sounding like a bulldozer with no engine sounds. Jon stood there in shock, his gun temporarily forgotten, and then he realized it was heading toward the house, the way his son had gone. He turned and ran to the trail, hoping to gain ground on it and cut it off before it reached the clearing. He hit the trail and ran as fast as he could toward the clearing, all the while hearing the creature thrash through the brush on his side.

He burst into the clearing and looked frantically about for his son. Jon Jr. was standing just inside of the fenced field, waiting for his dad. Jon screamed at him to run to the house, then he saw the thing crash out of the woods about fifty feet to his left. It crossed the ten-foot clearing and stepped over the fence in two strides, and was running through the field parallel to his son in a matter of seconds.

Jon screamed at his son to run faster, and took aim at the creature. He didn't fire because he was afraid to hit his son or his house, so he vaulted over the fence and ran in pursuit of them. He could see it angling toward his son, and knew there was no way his boy would make it to the gate before it cut him off. In desperation, he pointed the gun to the ground at his side and fired as he ran, hoping to scare it.

The creature veered more sharply toward his son, and put on an enormous burst of speed. He heard his boy scream as they seemed to collide; he saw the creature dip its shoulder down a little bit and suddenly Jon Jr. was airborne. He flew about ten feet then hit the ground rolling. It never paused, continuing to run at an amazing speed in a loop back toward the woods. Once the line of fire was clear, Jon stopped and squeezed off the remaining five rounds at the retreating creature.

He was pretty sure all the shots went wild; the creature never made a sound or slowed down, and was soon over the fence and back in the woods. He reached his son, who was shaken up but not physically hurt. He asked his dad if it was a bear. Apparently, little Jon was so busy running for the house that he didn't see the creature running after him. He said something big and black suddenly ran into him, he felt a huge paw hit his bottom and he said he felt like he was falling.

Jon pulled his son to his feet and they ran through the gate and into the house, locking the door behind him. They were both out of breath and white as ghosts, his wife screaming at him, demanding to know what the gunshots were for and if they were all right.

When he caught his breath, he told her to make sure the back door was locked; he was going to call the sheriff. He went to the phone and began to dial the number (this was before 911), then stopped and wondered what exactly he was going to say. He hung up the phone, realizing what an idiot he would look like if he told the sheriff the boogeyman just chased them out of the woods.

He told his wife that it was a large animal, possibly a bear. He didn't know how to begin to tell her their four-year-old was right; his cowman was real and it was more frightening than anything he could imagine. He told them all to keep the doors locked, and stay away from the windows.

Around ten o'clock that night, both boys were in bed, and Jon and his wife sat down to watch the news. They soon heard a loud moaning cry, kind of like the siren at the volunteer fire department. It would stretch out for a long time, and then end with a "whoop whoop" sound. It was coming from the woods opposite the house. His wife asked, "What the hell is that?" Jon answered truthfully, "That is Tim's cowman."

He then described to her the full details of what had happened, and she immediately wanted to call the sheriff. He persuaded her that they would sound crazy, and that he would handle it himself. She reluctantly agreed, and told him she didn't want either of the kids to go outside until this thing was gone. The howling went on until around midnight, when it got quiet again. Jon wanted to stay up through the night and watch over the house, but he had a long day at work and the excitement earlier had worn him out. They went to bed around one in the morning, and had no further problems that night.

They slept in that morning and the boys were already up and watching cartoons when they got out of bed. The first thing little Jon said was that he had heard the bear rubbing against the house last night. He said he was too scared to get up and tell his parents, and fell back asleep soon after. Then Tim said, "The cowman talks funny." This stopped Jon cold. He asked his son, "When did you talk to the cowman?" Tim replied "Last night, in my room."

Jon asked, "The cowman was in your room?" "No daddy, he's too big for my room. He talked to my window," Tim said, and turned back to the cartoons. "What did the cowman say, Tim?" Jon asked. "He talks funny; I don't know what he said. He talks like this—Ooh Ahh Ahh Ooh!" Tim said, and started making strange monkey-like noises.

"Did the cowman try to get in your window?" Jon asked, breaking out in a cold sweat. "He's too big for that. He made funny faces, he has Lincoln Log teeth!" Tim said with a smile. Jon later learned Tim meant it had square teeth that looked the same size as the small blocks in a Lincoln Log set.

It apparently spent quite a while "talking" and making

faces outside the boy's window. Tim said it laid down and went to sleep outside, and he could hear it snoring. Jon then walked to his younger son's room, and cautiously peered out the window. No sleeping cowman. Jon told the boys to get dressed; they were going to go visit their uncle in Elma for the day.

After his wife and kids left, he called one of the men from his crew, and asked him to come over. I'll call him "Patrick." He was an ex state patrolman and my dad said he was kicked off the force because of his drinking problem. He was a good worker and never got drunk before dark, so Jon figured they would have most of the day to look for this thing. When Patrick arrived, Jon greeted him at the door and said, "Are you up for some hunting?"

Seeing how it was not hunting season, Patrick told him he doesn't poach, and doesn't even want to know about it if Jon did. Jon told him it wasn't deer he was after, and went on to explain the previous night's events. Patrick didn't really believe him, but could see he was sincere and still shook up. Jon had his pistol and a bolt action 30.06, Patrick had a .38 in his car and Jon loaned him a 12-gauge. They first circled the house looking for any signs of a nocturnal visitor.

At the back of the house, there was a spigot for the garden hose, and it always leaked. There was a patch of ground worn bare of grass under it, and it had turned to mud. In the center of the mud, there was a huge, clear imprint of what looked like a bare, human footprint. Jon said it was at least eighteen inches long, and very wide. It was so clear that he got the feeling it was left there on purpose. They found no other prints around the house, and in places in the field and woods where a track could be made, the creature seemed to avoid them.

Off to the side of the track in the mud were four straight lines, about eight inches long. He said it looked like someone had raked their fingers through the mud. When they circled around the side of the house and got to Tim's window, they saw what it was for. Above the top of the window, a good seven feet up, were four muddy streaks. And on the window itself were dozens of large, muddy fingerprints. The glass wasn't cracked or broken, just smeared with mud. By this time, Patrick was fast becoming convinced something strange had indeed happened the night before.

Before going out into the woods, Jon wanted to feed the family's pigs. They had two of them, apparently fairly young, weighing around forty pounds each. The pig-pen was about a hundred yards away from the house, behind an old barn. As they got closer, Jon became concerned because he couldn't hear them making any noise. Usually they squealed like crazy when they knew food was near at hand, but this morning it was completely silent. They rounded the corner and the pen was empty. No sign of damage or struggle, the pigs were just gone. They searched the barn but found nothing out of place, so they decided to hit the woods and try to kill this thing.

They entered on the same trail Jon and Jon Jr. had used the day before; Jon showed Patrick the broken fence wire and told him again about the hair. It was a bright summer morning, and Jon was surprised at the difference from the previous evening. The night before had been still and silent, but now the woods were alive with birds and small animals. He showed Patrick the broken tree, and they followed the creatures' trail and found several more trees and large branches twisted and broken.

They could see large, faint impressions of footprints where the ground was soft. They followed the deer trail further into the woods, and encountered nothing unusual. By noon they were both getting hungry, so they hiked back to the house for lunch. They spent the rest of the day poking around, but saw nothing more out of the ordinary.

Just before dark that night, his wife and kids drove up. He and Patrick were sitting on the porch with their guns, watching the woods. His wife asked if they had seen anything, and Jon told her about the footprint and mud on the window. Patrick had retrieved a pint of booze from his car and was well on his way to getting smashed. Jon decided he didn't want a frightened drunk with a gun around his family, so he suggested that Patrick could go home. Patrick agreed and drove off, and Jon continued to watch the woods.

His wife brought out a plate of food, a Coleman lantern and a flashlight. He told her he would stay out there and watch the house through the night. Before they went to bed, he went into their bedroom, and with help from his wife, pushed the king-sized bed as far from the windows as they could. They agreed that his wife and kids would all sleep in that bed for the night, and he would keep watch around the house. She had grown up hunting and knew how to handle a gun as good as Jon, so she insisted on keeping the shotgun in the room with them. He agreed after making her promise to ask for a name before shooting anything. If it replied "Jon," please don't shoot it!

There was a full moon that night, and Jon could see across the field and into the inky dark of the woods. The night air was filled with the sound of thousands of crickets, and the pond behind the house was full of croaking frogs. As the moon rose higher, clumps of weeds in the field began casting sinister shadows, and before long, Jon was seeing big, hairy creatures sneaking up on him in each of them. He stood up and lit a cigarette, trying to shake the fear and concentrate on the task at hand. As he smoked, he wandered to the end of the porch, and stood looking at the darkened barn. Something was different, but he couldn't quite place it.

The front of the barn facing the house was open, and the moonlight was hitting it from the side, casting the interior in deep shadows. He stood watching the black opening as he finished his smoke, thinking about the missing pigs. He then realized what was wrong. All the crickets and frogs had gone silent. It was as quiet as the inside of a mausoleum at night; so quiet he could hear the minute shrill buzz of his own nervous system.

As he turned to walk back to his chair, he thought he saw movement in the barn. He looked intently at the opening and could make out nothing, then turned his head a bit to the side and saw what looked like two red eyes hovering about eight feet off the ground. He couldn't see them if he looked straight at them, but when he averted his eyes a little, they became clearer. They were a deep burning, coal red, almost invisible in the dark. Every few seconds they would disappear when the creature blinked.

His heart began thudding in his chest, and he waited for it to leave the barn and approach the house. He slowly backed up to his chair, never looking away, and picked up his 30.06. He walked back to the end of the porch and watched and waited. He stood looking at the blinking red eyes for what seemed like hours, and then the eyes blinked out and never came back. He watched intently but could see no movement.

He thought for a moment, then grabbed the flashlight and shone it at the barn. The flashlight was too small to penetrate the darkness of the barn from this distance; he had to get closer. He was none too keen about leaving the relative safety of the porch and confronting a glowing-eyed monster in his barn, but he was damned if he was going to live in fear in his own home.

He left the porch and began slowly working his way toward the barn, taking his time, and building his courage up. He got closer and could still see no movement; it had gone further into the dark. He got within twenty feet of the opening, and his flashlight would now penetrate the gloom in the barn. He moved the feeble beam of light over the contents of the barn, an old tractor, and old pickup, boxes and buckets.

Too many places for something to hide, even something big. He cautiously walked closer, now shining the flashlight down the barrel of his rifle. He stopped at the entrance and shined the light all over, searching the corners and under the vehicles. He stepped into the barn, every sense straining for sound or movement. He walked around the pickup, tensing for a huge, hairy arm to reach out and grab him at any second. He made his way clear to the rear of the barn without seeing anything, and slowly turned around to leave.

He felt both relieved not to have encountered it in the dark barn, and frightened and somewhat confused about where it could have gone. As he was walking out, he glanced at the wide stairs leading up into the hayloft and froze. He knew with complete certainty that it had climbed those stairs and was waiting for him to walk out under the hayloft and jump down upon him.

He couldn't move, literally frozen in fear. He swore he could hear the floorboards softly creak above him as an enormous weight edged stealthily closer to the edge. He stood with his heart pounding in his ears, unable to move or act. Suddenly, there was the booming explosion of a shotgun from the house, followed by his wife screaming. His paralysis broke and he bolted out of the barn toward the house, completely forgetting what may have been in the hayloft.

As he ran toward the house, he heard an inhuman roar coming from the woods behind the house. It sounded pissed off and in pain. It screamed again and he heard branches breaking as it plowed through the forest, thankfully away from the house. He got to the house and almost knocked down the front door in his hurry to get inside. He ran down the hall to their room and found his family huddled together on the bed, sobbing. One of the widows was blown out, and his wife was still pointing the shotgun at it. When he burst into the room she swung the gun in his direction and screamed, and he hit the floor. He waited for the blast but it didn't come. He slowly stood up and went to the bed.

He asked her what had happened, but she was too shook up to answer just then. Tim started crying, "Why did you shoot the cowman, Mommy? Why?" Jon Jr. had his face buried against her shoulder, crying. After they calmed down a bit, he told them to get up and follow him. He led them to the living room, then went out the open front door and looked carefully around. He could see no sign of the creature; all was quiet again. He told them to come out and get in the car. They ran out in their pajamas and piled in the car, and Jon got in and drove them to his brother's house in Elma.

On the way there, they had calmed down enough to tell him what happened. She said a couple of hours after they went to bed, she finally dozed off. She was awakened by Tim talking to someone, and this bizarre clicking chirping sound. Tim wasn't in the bed; he was standing in front of one of the windows. The moonlight was shining through both windows, illuminating the room pretty good, but there was a large shadow, like a tree obscuring the window in front of Tim. She knew there were no trees close enough to cast a shadow, so she told him to get away from the window. "Mommy, listen! The cowman can sound like a bird!" Tim said pointing excitedly at the dark figure in the window.

"Timmy, get away from the window," she said, trying to keep her voice quiet. Right after she spoke, the noises from outside changed. It went from a soft chirping, to a strange gibbering, almost like human speech with an occasional pig-like snort thrown in. At this time, little Jon woke up and said, "What is that?" rather loudly. This seemed to incite the creature and it hit the side of the house with its fists hard enough for the walls to tremble. At this, little Jon screamed and Tim yelled, "Quiet, you're going to scare him away!" She yelled at Tim to get away from the window again, and reached up on the headboard and grabbed the shotgun.

She got out of the bed and started toward Tim; the creature leaned down and looked straight in the window at her. She screamed and raised the shotgun, afraid to shoot because her son was so close to it. She started forward to grab Tim, and there was an explosion of breaking glass. A gigantic, hairy arm reached through the window toward her son.

She screamed again and fired over Tim's head, blowing out the rest of the window and hitting the creature with .00 buckshot. It jerked backward out of the window and disappeared into the dark. A few seconds later, she heard it screaming in the woods. "It was trying to get Tim, it was trying to grab my baby!" she started crying again and he comforted her as best he could while driving.

They stayed the rest of that night and the following night with his brother's family. He told his brother about it, but could see he didn't really believe him. He agreed to ride back to Jon's house with him early Monday morning before work. They had left the front door open in their haste to leave, and he was afraid animals or vandals would have got into the house.

When they arrived, the house looked like a tornado had gone through it. The couch was upside down. They had a large, heavy console TV and it was apparently thrown across the room, lying in a spray of broken glass. The kitchen was trashed; the refrigerator knocked over and food everywhere. The doors to both of the boy's rooms were left closed, and the rooms were untouched, same as the bathroom.

The master bedroom was torn apart, the pillows ripped up and feathers everywhere. The chest of drawers was knocked over and the large mirror smashed. Jon's brother looked around in awe, and said, "You better call the police!" Jon looked at him and said, "And tell them what? Bigfoot destroyed my house?"

They left and closed the front door this time, and drove to my dad's mill in Aberdeen. Jon's brother waited in the car while Jon went in and told this story to my dad. After he was done, my dad said, "Well, let's go have a look at it then." They drove back out to the house, and Jon showed my dad the damage. He pulled the clump of hair from his shirt pocket and let my dad look at it.

As they were walking through the house surveying the damage, my dad pointed out cracks in the ceiling where it had apparently stood up and hit its head. Jon told my dad that they couldn't live there anymore; even if the creature was gone, they would always be afraid.

Their homeowners insurance wouldn't cover the damage; the adjuster claimed Jon must have done it in a drunken rage. My dad helped them find a place in Aberdeen, and gave him a loan for new furniture and stuff. The house was eventually fixed up and sold, and my dad never heard about another problem there.

A few observations about this story: My dad lost contact with Jon and his family in the mid eighties. They moved out of state and my dad hasn't heard from them since. His brother died around the same time. Why didn't they call the cops? Jon had a lot of pride as well as a lot of common sense. He knew he couldn't logically explain what had happened to the authorities, and he didn't want the story to get out and have him branded a nut case. I asked my dad if they saved the hair, and he said that Jon never mentioned it again, and my dad never asked him about it.

I asked my dad if he saw the footprint and muddy fingerprints, and he said he did. He said it looked like a giant barefoot man had stepped very carefully in the center of the mud. He's not a tracker, but he said it was the clearest print of any kind he had ever seen. I asked my dad if the neighbors had heard any of this. He said if they did, none of them ever mentioned it. I also asked him if he thought it was possible Jon had made it all up. That he had trashed his house in a drunken rage, and made up this elaborate cover story.

My dad said Jon and his family were terrified of that place; they didn't even want to go back and get their clothes. If it was just an elaborate story, what did he stand to gain? To profit from a story in any way, you have to share it with people. My dad and the other folks mentioned in the story are the only ones who ever heard it, until now, of course.

He also said that whatever trashed that house was no man. The TV had to have weighed close to two hundred pounds, and it was obviously thrown across the room with great force. He said that even after two days, there was still a wild animal smell in the house. I asked him if he thought there might have been two creatures involved, considering the incident in the barn. He said he asked Jon that same question, and was told that Jon felt there was only one, that it had lured him into the barn, then snuck out the side door to the house. The thing he thought he heard in the hayloft was either his imagination, or some common animal like a raccoon.

For whatever reason, this critter seemed focused on their four-year-old son. Their son was the only one who never showed any fear of it. He seemed to think of it as his friend. And although the sex of the animal was never determined, it was referred to as a male because of the predatory, stalking type behavior. That and the conspicuous lack of breasts, or perhaps it was just not as well endowed as the Patterson film subject. Anyhow, its behavior almost seems indicative of a mother that has lost her little bigfoot and is looking for a replacement. I rather facetiously asked my dad if little Timmy was a particularly hairy child, perhaps suffering from that rare condition that causes uncontrollable hair growth all over the body. He said Timmy was a normal little boy, with normal brown hair on his normal head.

I didn't ask if Timmy regularly reeked of rotting garbage and wet dog; didn't seem a polite course for the conversation to take. He told me of other possible bigfoot encounters he and his crews had in the woods around Grays Harbor. None of them are quite as titillating as the "cow-man" story, but interesting nonetheless. Perhaps I'll share them if there is an interest here in them. So, in the end, I was left with no leads to follow, no new evidence of anything, but I did come away with a pretty damn good story.

The Lummi Reservation Bigfoot Visits, 1975

In the late fall of 1975, there were numerous (over one hundred) reports of sasquatch sightings on the Lummi Indian Reservation for several weeks. Huge tracks indicated at least three creatures, of different sizes, had been roaming the area.

One dark October night, approximately seven-thirty p.m., a Lummi Police sergeant responded to a prowler call at the home of seventy-eight year old Emma Smith. Arriving at the residence, the sergeant found two boards ripped from the smokehouse on the back property, and a storm door on the back of the house that was torn and its frame splintered, but no sign of the prowler.

At 2:30 a.m., another call came in reporting another disturbance in the back of Emma's home. The sergeant arrived, along with several other people. Shining the search light around, he spotted a huge, hairy creature standing in the brush. While someone else held the spotlight on it, the sergeant walked up to within thirty feet of the creature, which had crouched down but made no attempt to run. For "many minutes" they stared at each other, neither one making a move.

The creature was seven to eight feet tall and covered in short, black hair, except on the face. It had small eyes, a flat nose, and seemed to have no neck. It appeared to have four teeth larger than the others, two upper and two lower. The sergeant had a loaded shotgun with him, but did not want to kill the creature. They just stared at each other while he wondered what to do.

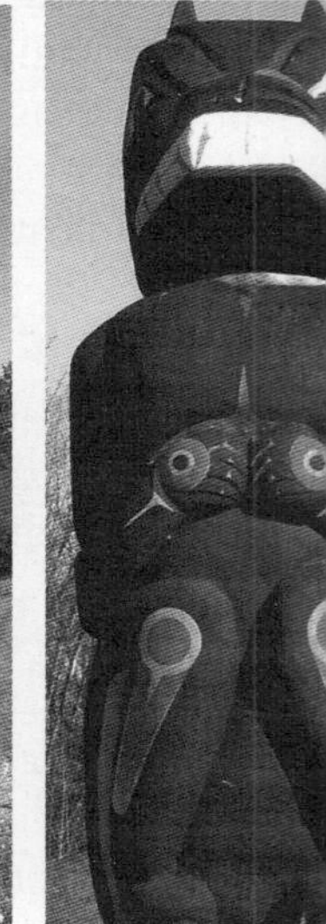

The Lummi Reservation border[s] the Pacific Ocean near the tow[n] of Ferndale, Washington. It is a beautiful setting, and comprise[s] Lummi Island, which has regular ferry service. Totem poles adorn the landscape, which make it a "world apart."

By now, seven people had shown up, including two more police officers. They could all see the creature, each one watching in amazement. Suddenly, loud noises came from the dark woods on both sides. The man with the spotlight swung it off to the right and called out, "There's another one over there!" At this point, the sergeant decided to leave and slowly backed away toward his patrol car.

At daylight, he returned to find many footprints on the ground. They measured eighteen inches long and seven inches wide.

Four days later, about 11:00 p.m., he saw one of the creatures again, seventy yards ahead of his patrol car. As he raced toward it, it ran across a field, bumped into two sleeping horses, spun around and raced into the woods.

The sergeant soon took another report from one fellow who was driving along a road when a bigfoot appeared, raced alongside the car at forty miles an hour for about 170 yards, before it leapt into some bushes.

The sergeant himself has heard the sounds the bigfoot make. He says they make several different noises—whistles, hoots, screams, and a noise like a bird. His brother, also a tribal officer, recorded what he says is a bigfoot scream. One day, he was playing the recording for about fifteen men out near the dump, when a bigfoot came running out of the trees, answering the screams. One of the guys took a shot at it and it ran back into the trees. At least three bigfoot terrorized the reservation for months, before eventually moving on.

The Green Mountain "Cover Up," 1980

The late researcher, Fred Bradshaw, once related a curious story about his father as follows:

In May of 1980, his father worked for Weyerhaeuser Company at Green Mountain, Washington, east of Vancouver. The site had security on their roads to check on the equipment and to keep people out of closed areas. His father was working the day Mt. St. Helens erupted. He was at a meeting in Kelso, Washington, and as he was a supervisor when the mountain blew its side out, all hell broke loose for him.

He was sent back to Green Mountain right away, but like most, he wasn't allowed to go very far because of the mudslide coming down the river. However, he did get to the town of Toutle, on Highway 504 off Interstate-5. He and his crew were placed at different spots to watch mudflow, and of course to help people get out of the blast zone after the first big eruption. He was then sent to the area of Spirit Lake to keep people out.

When the second major blow up of the mountain came, he and another worker reported in, and were told to get out of there. He was then placed in charge of the helicopter-landing zone. It was his job to help keep people out of the landing zone and let aid crews in so they could care for the injured.

Later, when all the people and the bodies were taken out, the National Guard was brought in to clean up. They hauled dead animals out that they had placed in piles; deer in one, elk in another, and so on. The carcasses were covered up with tarps and later doused with napalm and burned (napalm even burns the bones).

His father was placed in charge of one pile of dead animals in particular. The pile was covered and no one was allowed to come near it. Armed U.S. National Guard personnel were guarding this pile. On the day that they were to move this group of bodies, his father was standing very close to the pile and was told to keep his mouth closed about what he was to witness.

When the tarps were removed, he was amazed to see that the bodies were those of sasquatch. Some badly burned, and some not. They were placed in a large net and lifted into the back of a truck, which was then tarped over. His father asked a guardsman what would happen to the bodies, and the guardsman replied, "They'll study them or whatever. I don't want to know. It's like other stuff, you don't ask." Later that day, his father and the rest were debriefed, told not to talk about what they had witnessed there that day, and sent home.

Source: Ray Crowe, International Bigfoot Society — Report #3493.

Bigfoot and the Lost Boy, 1989

Harry Oakes is a tracking dog handler. He represents Mountain Wilderness Search Dogs, a search and rescue unit responsible for locating a multitude of lost people in the Pacific Northwest and around the world. His tracking dog, Ranger, had so far been responsible for finding eighty-eight people. On a Friday in July of 1989, Harry and Ranger got a call for help.

A three-year-old boy, Joseph Edwin Leffler, had disappeared from his home on Squaw Creek Road (near Delph Creek, South-

East of Estacada, Oregon (Clackamas County). He told his mother he was going "fishing" (play fishing) at a shallow creek, only two hundred feet from the house. The boy was lightly dressed, wore sneakers, and had his three dogs with him—two adult male dogs, both part dingo and part Australian shepherd, named Dan and Jack, and a small white puppy.

Harry and his search group showed up to search for the boy. About 6:00 p.m. on Saturday, Harry and Ranger found tracks of the boy and his dogs a mile west of the boy's home, heading downstream through marshy ground on nearby Delph Creek. Harry and his group thought somebody was playing a joke, as there was a series of barefoot, "four-toed" footprints, where the boy's tracks suddenly disappeared.

They could see where something described as "huge" had been walking through the brush. Harry measured one track where there was a good imprint (not sliding) at twenty-seven inches by ten inches, with a five foot stride (a twenty-four inch print has been attributed to a twelve-foot bigfoot). The track was a half-inch deep in pretty solid ground, and Harry said he stomped on the ground with his heavy boots and he "couldn't make a dent." Ranger sniffed; he ignored the bigfoot tracks, hunting for the boy's scent, which had mysteriously disappeared near a roadway.

The 939th Air Force Reserve Aerospace Rescue and Recovery Group had two helicopters with high-tech search equipment on board, including infra-red video cameras taping their progress while criss-crossing the area in a search grid. A report came in; one of the pilots flying up above stated that their FLIR system (Forward Looking Infra-red—a thermal imager camera) had picked up a huge object on the ground, "something they've never seen before." They requested Harry go and check it out.

Questioning the FLIR operator later, he said, "The thing was huge, upright and walking. Its physically impossible to be that huge on the system." He told the flight engineer to check it out. That's when they called Harry.

Looking around the area, Harry and his partner found more tracks. They were of the same huge size they had seen near the creek earlier. They also looked around for hair on the branches or scratches, such as a bear might leave, but never found anything. They did notice broken branches up pretty high—seven to eight feet up.

Just before noon on Sunday, in a wooded area near the search base camp, little Joey Leffler walked up to Judy Magill, one of the Mountain Wilderness Search Team members. "He stretched out his arms, and I picked him up," she said. "His first comment was that he was hungry."

When questioned later, Joey said a big, hairy monster had came and got him and the dogs and kept them safe. It had fed him berries and showed him where the creek was and how to drink. They slept in a mining cave. He said when the helicopter came the monster got scared and left. Beyond that, they were not able to question the boy as he was immediately helicoptered off to University Hospital in Portland, where he was checked and released about 7:30 p.m. The hospital said he was hungry and that his feet were wet and cold. There were no signs of hypothermia.

The boy's rescue and statement about the big hairy monster was taped on location by Channel 2 News. However, when the tape was aired they explained that the young boy told the story the only way he knew how, suggesting that it was only a tall tale. Other than what Channel 2 News reported, nothing was reported or mentioned in any other media source about the "hairy monster."

Harry interviewed one of the forest rangers about the possible bigfoot association and Joey. The Ranger said there are lots of mining caves up there, and lots of sightings had been reported, and that the forest service people were of the opinion that the thing had the IQ of a dog—not overly intelligent, but not stupid either. He also said it was not threatened by other animals.

It was later guessed that the bigfoot had found the lost boy, picked him up and carried him away. "We don't understand why the boy's dogs didn't object." A guess was that the dogs, dingo/Australian shepherds, being Australian in origin, didn't have the fear-hatred-kill instinct of other African/Eurasian dog types. The barkless dingos (maybe bigfoot don't mind quiet dogs) have been closely associated with the very primitive Aborigines. Some say they are survivors of prehistoric times (*The Dog,* F. Mery, 1968), and if the bigfoot critter is a "relict-human type," they might even be ancient allies.

Jack the Logger, 1998

(Submitted to Ray Crowe [via another researcher] by a witness who wishes to remain anonymous as explained in the story.)

> Back in 1998, along Blacksnake Ridge in the foothills of the Blue Mountains, I had an encounter with a bigfoot, and got a good, long look at him. Because of all the ridicule I've faced for leaking this story to people over the years, I'd rather not share my name. So just call me Jack the Logger, as I work in the logging industry. I was convinced to finally tell my story by two researchers named Vance Orchard and Bill Laughery. They told me the story needed telling, and I agreed.
>
> It happened on a cold day in late November of 1998. I make two or three trips per day when I'm hauling logs, and this counts a lot of days when I have to put on chains at the top, and take them off when I get out of snow. And, I could probably give a pretty good count of deer and elk in those parts as well.
>
> But these days, ever since that late November day, I've been eyeing the landscape for bigfoots! On my trip that afternoon off the mountain, I was headed toward the town of Dixie, a small town of a couple hundred people on U.S. Highway 12, and this day was the highlight for me of some forty-five years of trucking logs.
>
> I first spotted the creature when it was on an open slope some three-fourths of a mile away, but I didn't realize what the thing was. I estimated about where on the road I would likely cross its path. All that time, I'm wondering what kind of animal it was. It never entered my mind that it was going to be a bigfoot.
>
> As my rig came into the curve at the end of a long grade in the road, there was only about forty to forty-five yards separating it from me, and as I got there and saw him, I stopped my truck and shut off the motor. He was standing there in a heavy, tufted grassy area, just standing and looking at me. We both eyeballed each other real good.

I was close enough I could see his facial expressions. He didn't look like an ape in the face; more like man-features, but hairy in the face. I would say he had a nose, but not much. The skin was black and his hair color was like a smoky-gray, and had gray hairs showing like an old dog will get around his nose.

Anyway, while he was standing there, the expression on his face changed three or four times. That led me to believe that man may not be the only animal that has reasoning. That old boy was thinking and every time he'd go to a different train of thought, his expression would change. I was looking at the width of his shoulders and his height, wondering what the hell was going to happen!

He was a good yard or more through the shoulders and I've had people tell how a bigfoot is about eight-foot tall. Well, this dude was taller than eight feet and closer to nine feet tall. When you're that close, it's no problem to figure out how big it was. And he never made any effort to run from me. He never acted like he was scared. I sure know he wasn't scared of me, not a bit!

Then he turned and walked along with a limp, like something was wrong with one leg; like he had an old injury or someone had shot him. Then he stopped and turned and looked at me for another full minute before he left. He didn't run. He just walked over to the edge of the brush that dropped off steeply into the Dry Creek North Fork.

There was no getting around it, this was not any man-made object or a man dressed up; there isn't a man in this county big enough to wear that suit! This was my first bigfoot encounter, although several years ago I saw something that I thought was a bear standing up—always thought it was a bigfoot but could find no sign. But this time it's different, absolutely no doubt about it. I would pull fifty dollars out of my own pocket though, if one of you guys could've been there with me.

I didn't know whether to say anything to anyone about this—you know, if I'd go downtown and tell the guys I saw a bigfoot, they'd laugh me clear out of the place. I told my wife about it and she kind of had her doubts about it for while, but she knew I wasn't going to come in with some kind of cock-and-bull story to take a ridiculing over. But, I don't really care what people think. I just didn't talk about it, except with someone who has seen bigfoot or is a serious believer.

They can believe what they want, but I'm the one who knows what I saw. They can say there is no such thing, but they don't have anything to back that up, and I do. This thing was the closest to a real human than anything I've ever seen on television or real life. His body is proportioned more to a human than anything I've ever seen. He's not an ape. This dude walked like a man and somehow acts like a man. He walked like he was crippled in the right leg or foot. I'll tell you this much too, I've never seen anything like it, before or since. He's a one-of-a-kind for me!

Chapter 3

Eyewitness Reports

> Truth is like the sun. You can shut it out for a time, but it ain't going away.
>
> —Elvis Presley

The following entries are reports submitted to me by eyewitnesses. Many are certainly destined to become classics, and indeed predate some entries in this classification. However, because of my direct involvement, I have separated them. In some cases, the witness did not wish to be identified, so just his or her initials are shown.

I have classified the reports as either sightings, encounters, or incidents. Sightings indicate that there was no apparent interaction with the creature (it was simply seen). Encounters indicate that the creature was probably aware it was being observed or knew it was in the presence of humans and reacted in some way (but was not necessarily seen). Incidents indicate that the creature was not physically seen but, there was evidence of its presence (footprints, sounds, and so forth).

The Blue Mountains Bigfoot, 1947

Submitted by: Bill White, age 89, of Pendleton, Oregon (retired pilot, World War II veteran, Air Force. He flew B-25 bombers on fifty-six missions).

Sighting: June 1947, above the Blue Mountains of eastern Oregon (Union County).

> In 1947, I had not heard the stories about bigfoot, so I did not at first suspect that I had seen him. At the time of the sighting, I was flying a light airplane and skimming along an open mountaintop in the Blue Mountains of eastern

Bill White in 2007

Oregon. I was in the beginnings of my crop dusting business, an idea that was still new at the time. The farmers had gone into the pea growing business in a big way and they were plagued with pea weevil.

On that particular day, I was working out of Joseph, Oregon and was on my way back home to Pendleton. I liked to skim along as low as possible just to watch the wildlife. Mt. Emily was on my way and there was a long, open ridge with a jeep trail along it. That is where I saw the critter that I had not heard of before.

Off to my left a short distance I saw what I thought was a large bear (having nothing else to compare it to). But what was puzzling was, it was standing straight up on two feet. As I flew by, he turned his head and watched me go.

I didn't circle and look at him as I usually would because I was short of fuel and worried about getting to the old Pendleton airport.

If I had even heard of bigfoot I would have circled for another look. Makes no mind that I might have had to walk a mile or so home. There are farm fields a plenty, all of which are landing fields if needed. The next few days I told anyone who would listen (they were few) about the silly old bear standing up and watching me.

Over the years, the more descriptions I see and hear of the old boy's appearance, the more sure I am that the guy I saw was the real McCoy. When they describe how he stands with the broad shoulders, the long arms and the short neck etc. the better I can look back and see him standing there. That was no bear.

Note: Bill was a good friend and business partner of the late, well-known Kenneth Arnold, the pilot who coincidentally in the same June of 1947 spotted nine unidentified flying objects over Mt. Rainier. His sighting and report eventually set off the Project Blue

Book initiative. It was Arnold who first coined the term "flying saucers." White and Arnold were partners in the former's discovery of a large uranium deposit in New Mexico. The following newspaper article, which we believe appeared in a Pendleton, Oregon, newspaper in 1947, provides the complete story.

PENDLETON PILOT DISCOVERS URANIUM IN RHYOLITE TUFT

By Nolan Skiff

Carrol G. (Bill) White, 704 SW 13, Pendleton, commercial pilot, is the central figure in the first discovery of uranium ore of commercial value in rhyolite tuft in New Mexico.

The story was told here Friday by Kenneth Arnold, Boise, president of the Solar-X Uranium Corp., an Idaho corporation which is equal partners with White in the property.

White customarily spends some months during the winter near Demming, NM engaging in aerial crop dusting, then comes north for spraying work in this area. Last winter he purchased a scintillator and began aerial prospecting flights which eventually resulted in the remarkable discovery.

ALMOST INACCESSIBLE

The find was of approximately 160 acres in size in an almost inaccessible section of the Black Range mountains at an altitude of 7600 feet above sea level. He made the discovery by air Jan. 15 and then staked claims, later offering half interest to the Solar-X if it would develop and handle the property.

Arnold and his associates and engineers agreed to this after visiting and sampling the body of ore and obtaining assays at Moab, Utah.

Dr. Webber of the Socorro School of Mines at Socorro, NM also is much interested in the find from an academic aspect, since most big bodies of uranium have been located in limestone or similar formations.

CONFERS WITH AEC

Arnold has conferred with the Atomic Energy Commission through one of its officials, Glen Collins, Albuquerque, NM, and has been given encouragement. The ore is reportedly of relatively low grade but desirable in all probability to produce, although this latter will depend on the future decisions of the AEC. The commission already has assured the Solar-X Corporation that a crew of men would survey the property commencing about May 20.

Arnold said there was no way of knowing yet whether costs of mining the ore and transporting it to AEC receiving points would be too great to warrant actual operation, but so far indications are that it will prove itself financially feasible.

Bill White is seen here in his crop-dusting bi-plane.

Historical Note: *Bill White has led a highly interesting life. He sent me the following email, which I provide for your interest:*

It is possible that you might want the history of fire bombing. Here it is:

In 1958 Cal Butler of Butler aviation talked the US Forest service into trying airplanes for fire prevention. He took an old Navy torpedo bomber (TBM) and rigged it with a tank and set up a deal as to how it would be used.
It was known as initial attack. The pilot would get the first call on any smoke spotted and take out a load of borate (the substance used for the original suppressant). Low and behold they chose me as their first pilot for the experiment. I was stationed at John Day, Oregon. Sort of out of the way where they would not get too much flack in case it turned out to be a complete failure. However it was a complete success. Cal got them to permit him to put on a second plane in 1959 which was stationed at Lakeview, Oregon and he flew that one himself. From that year on we went into larger multi engine aircraft and other outfits got in to the act and thus the business grew.
Many changes as to how the planes would be used by the wise office personnel of the service has sort of ruined the effectiveness.

Bill

Art and Velma, 1958

Submitted by: Questa Knight, age 53, of White City, Oregon (real estate business).

Sighting: spring, 1958, Jacksonville, Oregon (Jackson County).

Back in the late spring of 1958, my parents, Art and Velma Backes, lived in the middle of Jacksonville, Oregon. I have heard their story many, many times over the years, and it never alters.

One dark and quiet night, approximately 11:30 p.m., they were driving home on the road into Jacksonville, where the orchard is now. Us two kids were asleep in the back seat. Suddenly, crossing the road directly in front of them bounded a large, dark, tall, hairy creature. It was approximately seven feet tall, ran upright, and took only four steps to cross the whole road. My father slammed on the brakes saying, "Did you see that?" My mother answered, "Oh my yes!" After that, not another word was uttered from either of them the rest of the way home.

My parents were very avid hunters all of their lives—deer, elk, bear, antelope, and duck. You name it, they hunted it. In 1958, they had never heard of bigfoot and had never seen an animal such as this in all their lives. They had no idea what it was. My dad was a log truck driver for forty plus years, in and out of many different forests, and had never seen such an animal.

They described the color as dark-brownish. They said its head was squatty and sat funny on the neck. Its arms were very long. And it was big, not fat, but big. And it had a strong stench to it. My mom always had a good nose; she could smell morel mushrooms and find them in the woods.

They had also, many times way out in the woods, ran across huge piles of scat, making them wonder, as no human could ever have left that much. Only something like the creature they saw could make a pile that large. My mom kept one huge pile, it was a medium brown color, and I would say it weighed three to four pounds. My mother recently passed and unfortunately someone threw it away.

Fishing the Stilly River, 1962

Submitted by: Reece Morris, age 61, of Granite Falls, Washington (U.S. Navy 1962 to 1966; retired metal fitter and metal artist).

Sighting: early summer, 1962 on the Stillagamish River, Granite Falls, Washington (Snohomish County).

It was back in 1962 and I was fishing the South fork of the Stilly River, north of Granite Falls and the old railroad tunnels. It was a beautiful sunny afternoon.

I was walking up river to find another fishing hole, when all of a sudden I got the feeling something was watching me, and the hair stood up on the back of my neck. I stopped and looked out across the river (there was a real high bank on the other side), and as I stood there looking all around me, I noticed some movement high up on that bank.

It was something big, tall and dark in color. At first glance I thought it was a bear, because I just got a glimpse of it. So I stood real still, trying to see if it would come back out into the open for a better view when all of a sudden it came into full view. It was not a bear.

This thing was standing up on two legs with its right arm extended, kind of leaning against the tree, moving its head back and forth as it looked up and down the river. And then it looked straight at me and I felt my heart coming out of my chest. It knew I saw it, and suddenly it took a step back into the trees and was out of sight. I got the hell out of there right then, 'cause I thought maybe it was coming down for me.

It scared the crap out of me; I was just a seventeen-year-old kid at that time. I tried to tell some of my family, but they thought I was just making it up. I'm sixty-one years old now and I still won't fish that area of the Stilly. They are out there. And you can print my name, to hell with those who don't believe it!

Rocks and Stumps, 1962

Submitted by: Steve Bray, age 60, of Elk, Washington (retired; Vietnam veteran, two tours; former commander of VFW Post 1474, Spokane, Washington). Many thanks to field researcher Linda Anderson.

Incidents: 1962 to present, Bead Lake, Washington (Pend Oreille County).

In the mountains of Pend Oreille County, in eastern Washington State, sits a group of lakes with names like Bead Lake, Cooks Lake, Marshal Lake, No Name, and Mystic Lake. All the lakes are within a three-mile radius of each other and close to the Idaho state line and Bonner County. In fact, when you take the road up to these lakes, you cross over from Washington to Idaho and back again into Washington, as the road winds along the river.

I grew up around these lakes and would explore these areas as a kid. I am very familiar with this country. In 1962, I was sixteen years old and living near Bead Lake. One late summer/early fall night, my family sat outside on the porch, talking and enjoying the warm evening. It was an unusually quiet night with no sounds of crickets, or any of the local wildlife, when suddenly, loud shrieks and screeches coming from somewhere off in the distance broke the calm silence. The shrieks sounded very close, but my dad said they were a good mile away. No one could identify the animal. I had never heard anything like it in my life. The chilling sounds went on all night long.

The next day I went off in the direction of the shrieks to see if I could find some tracks that might help identify what had made those sounds. I hiked about a mile up the ridge, never finding any tracks. What I found instead were some rocks, stacked neatly on a stump. I continued a search of the area and soon found another stump with another pile of rocks. And another. And another still. I found stumps with rock piles all throughout the mountain. I thought this odd, as this is a very secluded area. Yet the rocks didn't get there by themselves.

Examining the stumps, I found each had about six or seven rocks clumped on top, in no particular order. The rocks were various sizes, the largest being four to six inches and the smallest around two inches. Each one a different color (There are a lot of rich tones of rocks up there because the area is known for the granite beds). I took the rocks off the stumps, and by the next week they were back, neatly stacked on each stump. Each time I returned I would remove the rocks to see if they would wind up back there again—and they always did.

I got spooked one time while removing the rocks because I felt "something" was watching me. It was that "overwhelming feeling of being watched." I began to smell a pungent odor. It smelled real bad, like musty, wet dog, and dead animal. I quickly left the area and headed home.

The rock sites are all over the mountain ridge that connects Bead Lake to Marshal Lake. Very rough, dense terrain, mostly deer trails, then a little clearing where the stumps are. The brush is so dense you can see bear tunnels, around three feet tall. There are man-sized tunnels as well, not made by a man, yet made by something very tall.

To this day, and for many years now, a couple of times a year I return to the ridge to take the rocks off the stumps. And when I return they are always replaced.

My theory on the rocks: I think it is a possible communication with a family of bigfoot. To communicate where a food source is, or maybe where the best trail is. There are also tree crossings that appear to be barriers—so maybe a warning of some kind. Not necessarily to keep out, or to go away, but to warn others of a drop off or hazard.

Note: Bigfoot seems to have a thing about stacking rocks. Leaving stuff on stumps has been reported often. And often in response to gifts left there for them. Sometimes the "gift" left consists of rocks, spruce buds, pinecones, flowers, some little bones, and a dog skull in one case. A separate (non-related) type of rock stacking is the placing of larger rocks near game trails. The rocks are usually stacked about a foot high and are heated in the daytime sun. At night, small animals seeking the warmth of the rocks crawl under them. Bigfoot then collect the animals for food.

A Most Unusual Sight, 1965

Submitted by: John Wooten, age 56, of Portland, Oregon (Vietnam veteran, U.S. Marine Corps; automotive technician).

Sighting: late April 1965, Near the Chetco River, Oregon (Curry County).

When I was fifteen years old, my stepfather, my brother and I were out doing a little night hunting. It was on the secluded L&P logging road that runs along the Chetco River off Gardner Ridge Road, in Curry County, where we saw an extremely unusual sight.

Upon shooting a small forked horn deer and gutting it, a set of headlights appeared on the other side of the North Fork of the Chetco River. Because we were poaching, we thought it could be the game warden out to catch us. Knowing that he had at least five miles more to drive than we did in order to head us off at the pass, we put the deer behind a downed log and marked our spot on the road with a small branch of alder and started toward the cut-off where the L&P Road intersected with Gardner Ridge Road. We were traveling as fast as the old '58 Ford truck could safely go, when, upon rounding a sharp corner, there it was!

Moving along the road we were on was a most unusual sight. A creature, big and hairy like a bear, but standing upright and running like a man. It was at last six feet tall or better, big and broad and well proportioned. Its hair was a dark brown/tan color, and was coarse and matted. Its head was kind of triangle shaped, with not much of a neck. It continued to move across our field of vision, never once looking directly at us.

My stepfather exclaimed, "What the hell was that?" hit the brakes and the truck skidded to a stop. We watched it lope across the road for a good thirty seconds until it disappeared down into the gully. Upon getting out of the truck we noticed a rather strong, musky odor in the still night air. Our truck lights showed extremely large footprints where the beast had left the dirt road and headed down the gully. Not having much time to spare, we loaded back into the truck and continued to high-tail it toward the pass. Upon arriving home, we discussed the creature and all the "could've, should've, would'ves," and it's always been in the back of my mind that I can say I saw a bigfoot.

The Salmon River Sightings, 1966

Submitted by: John Elmore, age 54, of Vancouver, Washington (professional rock n' roll musician/bandleader).

First two sightings: Late August 1966 and July 1974, Salmon River, Northern California (Siskiyou County).

Growing up in the great Pacific Northwest, I have had several bigfoot encounters in my fifty-four years. My first experience took place in 1966 while on a camping trip to the Salmon River in Northern California. The Salmon River borders the Marble Mountain wilderness area, which is 5,000-plus acres with no access roads whatsoever. Hiking or going by horseback are the only ways to travel there.

I was thirteen years old when I went camping with my friend Mike and his parents. We set up camp at Jones Beach, a popular spot along the river, and then instantly headed off to swim and float our tubes. Mike and I decided to float down the river to the next bridge, a couple of miles downstream. After a mile or so we decided to head back to camp, so we floated over toward the bank and started our climb back up to the main road.

As we reached the top (a hundred yards straight up) we sat down and looked off across the river toward the other side. You could see where the flood of '64 had washed away everything for at least seventy-five to a hundred yards on the other side; there was nothing left but sand and dried mud.

We were scanning the tree line hoping maybe to see deer when we both noticed something moving out of the trees. I'm guessing that we were at least a hundred yards above it. We both looked at each other, "wow, a bear," yet knowing that a bear doesn't walk upright for any distance.

We were fascinated at what it could be. It surely didn't see us, or maybe it had watched us climbing the bank, but it didn't seem to be bothered as it slowly walked toward the river. As it got closer we could see it much clearer.

It was huge! I'm guessing between seven to eight feet tall and at least four to five hundred pounds. Brownish to black hair all over its body. It was hard to stay quiet, giggling to ourselves, even though we were scared shitless.

It slowly wandered on over to the riverbank, crouched down and started cupping water in its hands to drink. When finished, it stood up and began walking back the same way it came, toward the trees. By this time we had watched it for probably a good half-hour.

We jumped up and ran all the way back to camp; screaming to anyone who would listen that we had just seen a bigfoot. Being excitable young boys, we were told that what we probably saw was a bear. But we knew better!

***

My second sighting took place in the exact same area mentioned in my first encounter on the night of my twenty-first birthday party in July of 1974.

We had planned a kegger party down at the bridge below Jones Beach and invited about fifty-odd friends. Sometime just before 1:00 a.m. we were running low on beer, so we took up collections and three of us hopped into my car to make a run to town for fresh supplies. It was about twelve to fifteen miles back to Fort Jones, the closest town with a bar, so we were haulin' ass to get there before they closed. My old buddy Emmett, my wife Kathy, and myself.

I was driving as fast as I could on that crooked river road. As we came around the corner where the Jones Beach camping area is, I suddenly saw something in the middle of the road and slammed on the brakes.

As we came to a screeching halt, they both yelled at me, "What the hell you do that for?" I pointed up ahead and said, "Look there!" Sure enough, no more than twenty yards in front of us was a huge creature walking across the road, through the RV parking area and heading up the mountainside.

Now this had been the Fourth of July weekend, and dozens of people had been there just a day or so before. So most of the garbage cans, which were attached to chains, bolted to posts and cemented in the ground were full. To our amazement, we could see that this creature was carrying one of those fifty-gallon drums under its arm and dragging the chain connected to the post behind it.

It looked back at us once and kept on walking, crossing the lot and disappearing into the woods immediately. I'll never forget those eyes in the headlights; they had an eerie yellow tint to them.

After it was gone, no one said a word for several moments; then suddenly we all at the same time said, "Did you see that? Did you see that?" We burned rubber all the way to town after that. We hit the first bar we found to get our beer, and made sure to tell everyone in there what we had just seen. Some said, "Sure son, we know." Others looked at us with a face that said "So, what you been smoking?"

Beer in hand we headed back to the party, driving very slowly past the parking lot area at Jones Beach, just in case we might catch a glimpse of it again. We finally made it back to disgruntled drunks who wanted to know what took us so long. As we told them the story, and hearing some of the same remarks we'd heard in town, I told them, "Fine, there's proof!"

The next morning, we loaded up as many people as could fit in the back of a pickup and drove back to Jones Beach. Sure enough, you could see where the post had been ripped out of the ground, and the trail led straight across the road with that white concrete chalk mark on black asphalt.

We started to follow when one of the guys ran back to the truck and grabbed his rifle saying, "This is for just in case." We climbed about 200 yards or so up the mountainside and found the barrel, all smashed and dented in, with the remaining food and garbage strewn over about a fifty-foot area.

The place stunk to high heaven, a fragrance close to that of an open sewer and nobody wanted to stay there for very long after that. Needless to say, I got an apology from everyone there.

Third sighting: August 1979, Mt. Baker National Forest, Washington State.

***

My third experience took place in late August of 1979. My buddy Ray had been working for the forest service, doing fire watch in the Mt. Baker National Forest east of Bellingham/Sedro Woolley, in Washington State. He invited me to come up and spend the weekend with him out at his site. We had worked together in the past, so this would be a good time to catch up on what we'd been doing in between times.

No sooner than we got in the rig, we were swapping tales, and pretty soon he started in with his bigfoot stories. I had already told him some of mine years ago, so when he started in, I knew it was gonna be a whopper—whether he was bullshitting me or telling the truth. He told me that we would be staying in a fair sized camp trailer on a logging site landing, and that there was a family of bigfoot that lived in the area.

They didn't seem to want him there because they would come around every night after dark and throw rocks at the trailer. Now this was something I had only read about in stories, or heard tales that my father, who was a logger, used to tell about the strange things that happened at some of the places he worked at.

An hour or so later, we arrived at the landing, and one of the first things I noticed was an abundance of dents in the sides of the trailer, and many, many fist to football size rocks laying not far away. I said, "Damn, Ray, what'd you do to piss them off?" He just laughed and said, "Help me unload this plywood for the windows. Last time they were here, the bastards broke most of them."

This was beginning to get a little spooky already, and it wasn't even dark yet. So off we went, duct taping all the plywood up to cover the windows, and then grabbing the last of the supplies out of the truck to take inside: my sleeping bag, food, whiskey, another shotgun, and a couple more boxes of shells. Grinning, he looked at me and said, "You never know, they may get in here one of these times, but if they do, I'm taking a couple of them with me."

When finished with the windows, we settled in for the evening, listening to the radio and playing cards to pass the time. I was already spooked from what he had said earlier, so anytime the wind blew enough to rock the trailer I was asking, "What was that?" He'd just grin and say, "Go outside and check."

I was already scared shitless and nothing had even happened yet. "Don't worry," he said, "They might not even show up." We sat there for hours listening for every sound, sneaking a nip from the whiskey now and then (just for courage) until we finally dozed off a little.

No sooner than we were catching a short nap when—crack! crack!—we heard the sound of tree limbs breaking outside. A moment or two later, the first rock struck the side of the trailer. Wham! We were both up and clutching our shotguns in a heartbeat. "Shhhh, listen. I think there are several of them out there tonight."

We sat in total silence, listening to them grunting and calling to each other back and forth. Wham! – another rock would bounce off the trailer. Once in a while you'd swear they were rocking the trailer, as they would occasionally rub up against it.

Soon I began to catch a whiff of them. Gawd, what a stench! "Have you ever tried to scare them away?" I asked Ray. "Well, I have some flares," he said, "but that would require one or both of us to go outside, and as long as they're out there, I'm not going anywhere!"

So there we sat, all night long, listening to them wandering around outside, making strange noises and sometimes bouncing a few more rocks on the side or on top of the trailer. Somewhere around 5:00 a.m. it suddenly got quiet outside. We figured it was getting daylight and they had all went away.

Later that morning, we finally got brave enough to open the door and look around. Dozens of new rocks and what looked like hundreds of footprints in the soft dirt were all the way around the trailer. "Smile," said Ray. "You only have to go through this one more night."

"I'm afraid not," I told him, and caught a ride out with the first ranger making his rounds that day!

I told the ranger what had happened. He said that Ray was one of many who had reported the same experiences and that it was very difficult to get anyone to stay up there for very long.

Backyard Camping, 1970

Submitted by: Kelli Miller, age 48, of Portland, Oregon (mortgage banker, sales manager, writer).

Sighting: July 1970, Morton, Washington (Lewis County).

My story is from my childhood. As a young girl, I was raised in the foothills of the lower region of the Cascade Mountains, in a small town called Morton. Stories of bigfoot surrounded us and we repeated most of them at sleepovers and campfires. Such delight we took in sharing the tales of the tall, hairy creature. We didn't care if they were true or not. Little did I realize they were. And had I realized this, I wouldn't have been camping out with my friend in her backyard that night when we came upon not only his grunts and groans, but also a quick glimpse of him.

My girlfriend, Patti, lived in the country on the outskirts of Morton. Her family of six had encountered numerous visits from hobos and train-track tramps throughout the years, so my apprehension to sleep outdoors was a bit high. But after being consoled by Patti's mom that hobos only came around in the daytime for food, I felt more at ease and agreed to pitch the tent in her backyard, which faced numerous acres of wilderness. Her father started a small campfire for us, so we could enjoy a little warmth on that cool July night. We placed our sleeping bags outside the tent, crawled inside them and sat by the fire.

We began to share our boyfriend stories in between bites of burnt marshmallow, and snickered and snorted at quite a few of our silly little thoughts. "Joey had a tongue like a lizard," or "yeah, but that's better than being kissed by fish lips," and so on. It was somewhere in our laughter and the darkness of the night that I felt I was being watched. It was a very spooky feeling. I pulled myself deeper into the sleeping bag and we continued the stories as the fire simmered down.

A howling began in the distance. Coyotes. My eyes were as big as the full moon at this point. I searched the woods for any sign of danger. Nothing. Yet, there was something out there. I heard a few branches snap, one after another, like someone or something's heavy footsteps. Patti was still talking. I hushed her with my fingers and nodded toward the woods. She of course rolled her eyes at me.

A grunt, almost a snort, came from deep in the wilderness. Patti now quickly turned her face to view the woods, which were behind her. I was frozen. She said, "A deer, we always get them around here." We kept watching and listening for more snaps of twigs and branches.

Then, with the cinders of ash still glowing red, and the brightness of the full moon, we saw him. A tall, dark figure, maybe over seven feet tall! Huge! His eyes caught the reflection of the moon; they were intense, golden. His body was covered with hair; it was dark and difficult to see the color or thickness. His arms seemed to be unusually long.

This great beast seemed to acknowledge our attention and slowly stepped back from the front of the tree line and into the shadows. Patti and I were petrified at this point. Neither one of us could move or scream or say anything. I opted to scream but there was no voice in my throat. This is what you call "scared shitless!" A few more grunts and then the sound of tree limbs being pulled down. The smashing branches and the groaning faded into the distance, as well as the heavy, thundering footsteps. Soon, all was silent again.

I'm not sure which one of us gathered ourselves and ran back into the house first, but I do know what we saw and heard, and now understand it was a bigfoot, watching us from a distance.

The Fort Lewis Sighting, 1970

Submitted by: Steve Bray, age 60, of Elk, Washington (retired; Vietnam veteran, two tours, former commander of VFW Post 1474, Spokane, Washington). Many thanks to field researcher Linda Anderson.

Sighting: early spring 1970, Ft. Lewis, Washington (Pierce County).

> One dark night in 1970, I was assigned duty at the ammunition dump used for Ft. Lewis and McChord Air Force Base. It was the early seventies when a lot of anti-government militants would steal guns and ammunition from the bases for their cause, so the ammunition dump was heavily patrolled. There were six other guards on duty that night, and we rotated every four hours to keep alert.
>
> A ten-foot-high cyclone fence, netted with barbwire and four rows of barbwire on top, surrounded the ammunition dump. The outside surrounding land was guarded only with cattle barbwire four feet high.
>
> The night was chilly and very dark. The only light was coming from McChord Air force Base off in the distance. I was sitting in the guardhouse and would periodically monitor the other guards who were walking around the perimeter with walkie-talkies. They were to report to me if they saw anything unusual.
>
> One of the guards, a young man of about nineteen, was walking his perimeter about 150 yards from the guardhouse. He was walking 100 to 150 yards back and forth along the fence surrounding the ammo dump. About four feet from the fence ran a road. Only half an hour into our shift, the young guard radioed to report what appeared to be the very tall silhouette of a man standing across the road from him. I radioed back telling him to stay alert and to report back, but no sooner had I spoke than the young guard showed up in the guardhouse with me.
>
> He was hysterical and rambling on that it wasn't a man. He told how it instantly crossed the road in three strides and

was on him in a matter of seconds, coming to a stop two feet in front of him. He looked up to this huge creature towering over him, saying he had never seen anything like it in his life. But he knew what it was; it was a bigfoot!

I was more shocked that the soldier had left his post and showed up without his gun and radio. The soldier, realizing what stood before him, had panicked, dropping his weapon and radio. I saw how shook up he was, but thought it must be militia out there pulling a trick to gain control of the young guard's gun and the ammo dump.

I notified the sergeant on guard about the incident and he, another guard, and I went back to the scene and retrieved the young guard's gun and radio. He refused to go. While there, I smelled a strong, foul, musty skunk smell, and spotted several large foot prints in the damp soil by the fence, approximately seventeen to eighteen inches long. I also saw clumps of hair stuck to the barbwire on the cyclone fence. The hair was dark and stringy, about four to five inches in length. I knew then that the guard was telling the truth, and I was a bit tripped out that the thing might still be nearby.

The young soldier refused to stand post again, so the other guard stayed to take over his post. When I returned to the guardhouse, I reprimanded him for throwing his gun and radio down (but couldn't blame him). The soldier finished his shift, but after that night refused anymore guard duty.

The sergeant on guard had called the CID (the Army Criminal Investigation Department) and the Military Police. The Military Police were there within ten minutes, with M-60 machine guns mounted to their jeeps. Three jeeps, three men in each jeep. The CID showed up soon after in sedans. Two in suits, the others in street clothes. We were advised by CID to stay in the guardhouse and not to be at the site while they were working, so as not to contaminate the crime scene. It was treated as a crime scene because of the militant threat. We had to fill out a report and were told by CID to treat this as a routine investigation and to report "suspected militant," not "bigfoot."

It was dark, but from a distance, I did see them taking measurements of the footprints and what I thought to be casts. The sergeant of the guard later told me that they did plaster several casts and collected the hair samples. There was never any follow-up questions asked of me after that night, and I never heard another thing about the incident.

Note: Fort Lewis is the largest military base in the country. Located on the southeast edge of Puget Sound in western Washington, the military reservation constitutes one of the largest remaining intact forest areas in the Puget Sound basin, consisting of 84,000 acres of forests and prairie land cut from the glacier-flattened Nisqually Plain. Fort Lewis has a long history of bigfoot activity.

Horse Sense, 1970

Submitted by: J.R.C., age 57, of Amboy, Washington (regional manager for a pet warehouse in Canada).

Incident: summer 1970, the Skyline area, off Rocky Point Road, northwest Portland, Oregon (Multnomah County).

One late afternoon, my sister Sue and I were out riding our horses. It was getting late, and we didn't want to be on the pavement after dark, so we decided to take a shortcut home, along an old trail from Highway 30 to Skyline Drive. The secluded, five-mile steep terrain trail traveled right up the side of a hill, sixty-degree grade, and was very timbered with forested old growth. It was a well-defined deer trail, not very wide and brushy. Only single-file horses. You could easily miss it if you did not know it was there. In fact, because of the grade of the land, no one went up there much at all.

Traveling along the trail I noticed how very quiet it

was, no birds chirping, no noises at all, other than what noise the horses made. It was a creepy quiet. We had gone about three and a half miles in on the trail when the horses suddenly stopped dead (I almost went over the front) and they became extremely nervous. Both horses whirled around and around like they were confused. They became very panicked and hard to control—eyes wide, snorting—but the real problem was the whirling around in a circle. These were experienced trail horses, and we were both experienced horsewomen. We had even done trail class at shows with them, so they had seen and were familiar with bear, deer, and cougar. Nothing really bothered them, and never anything like this!

Then we smelled this horrible odor, which was overwhelming, something I have never smelled before or since. We never saw anything, but whatever it was, was close, because the horses just wanted to run. So, being the chickens we were, my sister and I did just that, let them run! Racing straight up the hill until we got to the road and then headed straight for home. Later, Sue said she felt something was shadowing us for a couple of miles. She kept seeing things out of the corner of her eye, yet nothing definitive, only a feeling we were being watched.

We didn't tell our family right away; instead we went to see an old friend, who lived at the bottom of the same hill. When we were kids he had told us of an encounter with a bigfoot in his pasture, in that very same area. He described it as hairy, not human, and very large. He was a six-foot, six-inch guy, and it was much bigger than he was. Talking with him helped confirm our suspicions.

Around that same trail, at different times, there were many reports of that same smell, in different areas within the same area. Neither one of us could ever describe it; we have tried on many occasions. We never went anywhere near that trail again. It has since been all developed and whatever was there is gone now, along with the trail and the forest.

Strange Incidents at Spirit Lake, 1970

Submitted by: J.C.C., age 51, of Mt. Vernon, Washington (janitorial services worker).

Incident: July 1970, Spirit Lake area of Mt. St. Helens, Washington (Cowlitz County).

My Uncle Gary was an avid fisherman and outdoorsman, who one summer invited me to accompany him on a fishing trip to the lakes above Mt. St. Helens. The lakes are all gone now, although Spirit Lake is trying to make a comeback. At this time in 1970, however, all the lakes flourished.

Upon arriving at the mountain, we parked in the lot next to the Spirit Lake store. Harry Truman was the owner of the store and boat rental (see notes below). When we entered the store, we were greeted by Harry. I remember he was wearing a flannel shirt with bib overalls. He really loved to talk about the lake, and he invited us out to the dock to sit and have coffee while he rambled on about the lake and where to find the trails to the ponds above.

He even offered to take us across the lake to the far side in his boat. We were there to hike, so we had to turn him down. After coffee, he ushered us back inside the store to show us some worms he thought would work great for the fishing. We bought two dozen of his worms and a bait carrier that could be attached to a belt. (Note: I was really saddened to hear he wouldn't leave his mountain before it erupted, in 1980. I was in the military at the time and out of the country. Harry really left an impression on me as a happy-go-lucky guy who was a great talker. He was someone who could talk with you like he knew you forever.)

We then proceeded to hike with full packs and poles up the trail. The trail went around the lake to the far side, and from there you could take two trails; one went up the mountain and the other meandered north away from the lake. We took the trail that meandered around the north side as Harry mentioned some beaver ponds that we could fish if we didn't mind the hike.

We had hiked all day when we decided to pull off the trail and make camp for the night. It was still light out when we set up our small two-man tent and built a small campfire. After a meal of hot dogs and beans we made coffee and sat around and talked. It was a nice, warm summer night with a full moon, and I remember discussing how red the sky was that night. We turned in around 11:00 p.m.

At some point during the night my uncle woke me, shushing me to listen. We heard what we thought might be a bear messing around the campsite. There were a lot of very loud noises and grunting sounds that went on for a good five minutes or so. Instantly the noises stopped, and the creature seemed to stand quietly still for a few minutes.

Suddenly, the grunting began again, only harder and deeper this time, and suddenly a loud thud next to the tent that literally shook the ground. We then heard (and felt) the heavy footsteps in fast stride running off into the woods.

We lay in the tent for about thirty minutes, afraid that whatever was out there would return. Finally we were able to pull ourselves together enough to talk each other into moving. We took our old nine-volt lantern and crawled out to investigate, only to discover a large tree laying lengthwise directly alongside our tent. It had been pushed over, roots and all.

Had it fell on our tent, we would have been seriously hurt. We checked the tree; it was not dead and appeared to be very solid, and we found no claw marks on it as a bear might leave. We had camped in a mossy type of area and as we looked around, we could see the ground was torn up real good as was the side of the hill the creature had obviously went up when it left. We found no tracks of any kind in the moss. But there was a smell, a bad smell of decay. At the time I thought my uncle had crapped his pants.

We didn't stay after that. We were so scared; we packed up in the dark and returned to the parking lot to sleep in the back of the truck. As we hiked back, I made a comment about my uncle crapping his pants back there. He laughed and said he thought it was me!

The grunting we later surmised were the grunts from pushing the tree over. I believe the tree was a warning that we didn't belong there. I do not believe it was a bear as the tree just wasn't the size a bear could push over, and the partial can of beans that were left out was not touched. The grunts were much deeper than a bear, and the bad smell of decay was a distinct smell that I've never smelled before or since.

At the time we thought if we didn't leave we would never make it out alive. I now believe that if it really wanted to hurt us, it would have. I never did return to the area, although I don't think it was because of the encounter. I now spend time fishing and camping the North Cascades, and keep aware of my surroundings. I am a true believer of bigfoot.

Note: To many, Harry Truman was *Mt. St. Helens and Spirit Lake. Not to be confused with the late president, Harry S. Truman, this Harry owned and operated the Mt. St. Helens Lodge at Spirit Lake, located at the north base of the mountain. He was a colorful, cantankerous, yet lovable old man. Born in West Virginia in 1896, he served our country in World War I and was a bootlegger during prohibition. He brought his family to Spirit Lake in 1926 where he started logging, hunting, and farming.*

The main lodge was built in 1939 and was very popular. Activities included boating, fishing, swimming, camping and hiking. After his wife passed away he lived at the resort with his sixteen cats. In 1980, Mt. St. Helens began acting up. In the evacuations that took place before the eruption, Harry refused to leave his lodge and became popular with the press. He didn't believe there was enough material in the mountain to kill him. He once said, "If I left this place it would worry me to death!" Harry was eighty-four years old when he perished, along with his cats, in the eruption on May 18, 1980.

Window Creature, 1970

Submitted by: T.L.W., age 40, of Medford, Oregon (business owner).

Sighting: summer 1970, Central Point, Oregon (Jackson County).

In 1970, my family moved from California to Jackson County, Oregon. My dad bought a "fixer upper" on Old Stage road in Central Point. The house had been vacated for at least six months. It was a one-bedroom house on four acres.

My mom and dad slept in the living room on a hide-a-bed, and I slept in the same room as my two older sisters. There was a door in our room with a window, and another window on the other side of the room that looked out to the driveway. There were no curtains on either of the windows at the time.

It was a calm, mid-summer night when we went to bed. At some point during the middle of the night, there was a noise loud enough to wake me out of a dead sleep. I looked toward the door and saw something big and hairy walking by. Just then it turned to look in, its face appearing in full view as it made eye contact with me.

Its face, which was surrounded by long, dark brown hair, looked like an ape. Its eyes appeared to be red. It must have been close to seven feet tall. Frightened, I yelled as I ran to my sister's bed. It quickly walked around the house toward the other window, where I saw its hair brush against the glass. Waking my sister, she looked to see a hairy hand on the window. I covered my head with her blankets, begging her to see if it was still out there. She finally looked, just soon enough to see the big hairy creature running off with big strides down the driveway.

I went into my parent's room and woke them. They said it was most likely a deer that I saw. There was no way this creature was a deer. I know what I saw, and it has stayed with me all of these years. The next day, I looked for footprints and evidence outside. The ground was rock hard so I didn't find anything.

I was frightened that night, but I don't remember feeling threatened about what was outside. I know it was just as startled to see me, as I was to see it.

There were many reports of bigfoot sightings in the area for the next two years, one in a gravel pit just up the road from us, on Old Military Road.

Two in a Lifetime, 1971 and 1998

Submitted by: Karen Bitz, age 51, of Molalla, Oregon (truck driver; Special Olympics coach).

First sighting: summer 1971, seven miles south of Oregon City, Oregon (Clackamas Country).

I have seen two different bigfoot in my lifetime. The first one in 1971, when I was in high school at our farm located seven miles south of Oregon City, Oregon (between Canby, Carus and Oregon City).

One late summer afternoon, my brother and I were walking out to the far barn on the back of our property, to feed our beef cattle. We had just walked past the first barn closest to our house when we heard a ruckus coming from inside. We stopped in our tracks, scared to death, knowing that dad and mom were both at work, and our other brother and sister were back in the farmhouse.

We stood there listening and watching when suddenly a tall, white-haired creature came walking out of the barn, heading across the pasture toward the canyon south of our farm. We were so scared that we dropped to the ground, not knowing what it might do if it saw us.

The bigfoot was all white, a pure creamy white color, its hair about four inches long—hair everywhere except its face. It walked upright like a man with long arms swinging back and forth. It was tall, maybe seven foot. It was not fat;

it was trim and muscular. It headed straight down the path from the barn. We watched as it walked across the back ten acres of pasture to the wooded area toward the canyon behind our property.

It walked at a normal pace, not rushed and not slow, long strides with arms swinging back and forth from front to back; walking like it was just checking things out. It did not appear to be old; it was moving and acting like a young man in good physical shape.

We were scared spitless, and as soon as it was no longer in sight, we ran as fast as we could for the house, locked the doors and called dad to tell him what had happened and that we needed him to come home—NOW.

When our dad arrived, he went out to check around the barn area, and said that the creature had bumped over some of the feed buckets and pushed off a lid from the barrels of cattle and horse feed. We stored our livestock pellets in fifty-five gallon drums at that time, behind the mangers. There were no footprints to be found as it was summer time and the ground was dry and hard.

Second sighting: February 1998, Banff National Forest, Canada.

The next time I saw a bigfoot (about thirty years later) was several years ago when my dad and I were both transporting RV trailers to Calgary, Alberta Canada.

We had reached the Banff National Forest park area, west of Calgary, where it was snowing lightly. It was very early in the morning and dad was driving his truck and trailer in front of me. I was directly behind him with my truck and trailer, when we saw a dark object start to cross the road from our left in front of us.

My first thought was a black bear or elk, which is very common in that area of the wilderness. But this animal was dark and hairy like a bear, yet walked upright like a man. It was about six feet tall and covered in dark brown-black hair, about four inches long, except for its face and hands. It was carrying a small, hairy figure in its arms, like a mother carrying a small baby cradled in her arms.

The creature continued to walk across the road in front of us, glancing in our direction like it was scared of our headlights. We slowed down, not wanting to hit it. As soon as it got to the right side of the road it started walking up the bank. She *[it was now thought to be a female]* glanced back at us and then tripped and dropped what she had cradled in her arms. She carefully picked it up, and with it back in her arms, she continued climbing up the bank.

My dad pulled over to the side of the road and came walking back to my truck window; his eyes were big and he looked shocked. He said, "What did you see? Did you see what I saw?" (He wanted to know it was real, that he wasn't crazy and seeing things that weren't there). I told him, "Yes, I saw it too. It looked like a bigfoot." He was excited, amazed, and dumfounded all at the same time. He told a lot of people about our sighting when we got back home to Oregon.

Both creatures I saw looked similar to the *Harry and the Hendersons* movie creature. Both had similar facial features and walked upright, like a man. Both rather tall, one had white hair and the other was a dark brown-black color, and both were in areas away from humans. I know in the Banff Forest Park area there was NO civilization anywhere around. Both encounters will be forever on my mind. I definitely know there are bigfoot creatures.

Karen's Father's Account of the Sighting

It was twenty-seven degrees (we have computers on our trucks with the temp), and two inches of fresh snow on the ground. We were driving trucks loaded with new RVs to be delivered to Red Deer, Alberta, Canada. Ours were the only tire tracks on the road.

We had crossed from Cranbrook into the Kootenay NP; we were about thirty miles from Banff National Forest Park. At 3:00 a.m., I noticed something moving on the left side of the road. As I approached the same spot, I saw in my headlights an upright creature that crossed the road in front of me, and was now on the right side.

It looked at me as if my headlights had frightened it. The creature then fell to its knees, dropping something as it did. What it dropped appeared to be a bundle wrapped in a piece of dark tarp, about a two by three-foot piece, with a two-inch piece of reflector tape on it.

I guessed that it had been taken from a construction site. But this was way out in the wilderness. The creature stood, quickly picked up the bundle and holding it closely to its chest, strode rapidly up the bank and away from the road.

I guessed that it might have been carrying a baby wrapped in that canvas, by the way it cradled the bundle so tightly to its chest. It looked back over its right shoulder, directly at me, and then was lost in the darkness. The creature was about six feet tall.

I could see no breasts (as it was holding the dark bundle). It had a full body of brown-black hair. I had a full-face view of the creature. The face was smooth and bare (no hair), the rest of the body was fully covered with about four-inch hair. I was amazed and speechless at what I had just experienced with my own eyes.

Tulalip Indian Reservation, 1972

Submitted by: C.R.S., age 52, of Tulalip, Washington (surveillance observer in gaming).

Encounter: October 1972, Tulalip Indian Reservation, Washington (Snohomish County).

It was an early afternoon in October when my two buddies and I caught a ride out to Tulalip Shores Road to go hunt an area that had just been logged off from Marine Drive down to the beach, about a mile long and a half mile wide. Most areas were cleared and some had patches of timber and brush, an area that is today all grown over.

We hiked quite a ways, from Marine Drive all the way down to the edge of the bluff by the beach. On the way back we discovered a small lake; at this point we decided to split up and go our separate ways, meeting later somewhere in the middle.

After an uneventful hike, I met up with one buddy. As we stood talking and smoking, we heard tree limbs breaking in a patch of woods near the Marine Drive hillside. Whatever it was, was big and heavy and the sound of limbs breaking became louder. We thought it was a bear, so decided to load our rifles. I had a .22, my friend had a .303 mouser, and the other buddy who was still out there had a 30:30.

We loaded up and began heading toward the sounds just as our buddy came running up to us, obviously upset. He kept saying we had to get out of there. We asked him what was wrong and told him we heard a bear right here in the woods, when he began to yell that it was not a bear.

"We have to go and we have to go now!" and he took off running. My friend and I looked at each other and decided that maybe we better go too and took off running after him. After reaching Tulalip Shores Road, he calmed down enough to tell us again that it was not a bear.

We told him it had to be and began to laugh at him

when he began to yell that bears do not walk on two feet! "I don't know what it was, I never saw anything like it before. You should've seen how big it was! It must have been at least ten feet tall and had hair all over it. It walked all the way across the landing, just like that, when it took us about fifteen minutes to walk it. And it went up that steep hill to Marine Drive in only a few steps. As it went up the hill it swung its hand and broke a tree right in half. I laid down when I spotted it so it didn't see me, but it was looking down at you guys standing where you were."

That's when we decided it would be dark in about an hour and we'd better get going and quickly headed out. As we walked down Marine Drive, we found the place where my buddy said it came up the hill. He pointed out the tree it had hit with its hand, which, just like he said, was broke clean in half. Not cut, but like when you break branches off a tree. We then found several large footprints by the side of the road, coming up from the landing.

We compared the tracks to our own feet finding them to be about five inches longer than our own and twice as wide with shoes on. We could see where it traveled up to the road and continued up the hillside on the other side. "Let's go, let's get out of here."

The area we had to walk was, and still is, a desolate area of road, about two miles in length. We had about a four-mile walk to reach home (which was close to the current Tulalip fire station) and there was no traffic for us to hitch a ride with.

Just before reaching the top of the hill, we began to hear soft footsteps in the woods that bordered the sides of the road. Footsteps that sounded like someone trying to be quiet in the brush, but the brush was dry, therefore we could hear them. Whatever it was, it was following us.

As we hurried along it began to make a lot more noise, breaking limbs and making loud stomping sounds with each step it took. Every now and then we would get a strong whiff of something bad; I don't know how to describe it now, just a very bad smell. We picked up the pace and hurried along, without running. We were always taught never to run from anything that scares you. So we walked as fast as we could.

About a quarter of the way down the long road, we came to the gravel pit, a place where everyone in the area used to go shooting their guns. At that time it was open and clear (it's all grown over now, but still there). The pit was maybe 200- by 200-foot round and surrounded by a high wall maybe thirty to forty feet high. We figured if we go stand in the middle of the pit that it might not follow us there, so we went and stood in the center.

Before long, the sound of limbs breaking and loud footsteps began again, this time echoing from all around the top edge of the gravel pit. After several minutes of loud noises, everything suddenly went quiet. Very quiet. Too quiet. And by now we didn't have much daylight left.

For a few seconds we thought, "All right, it gave up, it's gone," when suddenly gravel and rocks began sliding off of the top edge of the pit. We couldn't see it, but knew it was above us walking around the top edge. We put our backs together and raised our rifles; we shot a few rounds around the pit, hoping to scare off whatever it was. The rocks continued to fall when we realized we were out of shells; we had wasted them without evenfazing the thing, as well as losing more of our daylight.

I had my sheath knife and one buddy had his big pocket knife, but we were out of shells. By this time, we figured we were in trouble and decided to get moving before total darkness fell and we wouldn't be able to see anything. So we began to slowly walk out to the pit entrance.

Two cars had passed out on the main road while we waited in the pit, so we had already missed a possible ride. Halfway out, my friends stopped, thinking it could be waiting for us out by the road; they didn't want to leave the pit as it was an open area and we could see farther. I said, "If we stay it will get us, so we have to go now." I took off alone and wound up walking down the middle of the road with my hand on my sheath knife. My friends soon followed.

I was maybe 100 feet in front of my buddies when I began to hear the footsteps again. I could not tell where the steps were coming from. The sounds seemed to be echoing; I literally could not tell if it was on one side of the road or the other. I slowed down my walk to better hear where the steps were coming from when they began to get louder and louder. Suddenly, I heard a limb break while at the same time from behind, I hear my buddy say, "There it is!"

I looked in the direction of the noise and stopped frozen in my tracks. There in the brush, not more than fifteen to twenty feet away stood a creature. It was just standing there looking at me. We made direct eye contact. It was about nine or ten feet tall. Its entire body was covered in dirty white hair about four or five inches long—except for the face, the hair was much shorter on the face. And there was no hair on the palms of its hands. Its body was shaped like a human, but it was very big and very wide. And it was not human.

I put my hand on my sheath knife, not knowing what it was going to do. The creature just stood there, not moving, only its eyes and head following me as I slowly continued to walk down the road. It was a very intense moment. Suddenly my buddies went zipping past me. I hollered at them, "Hey, you're not supposed to run from these kind of things!" I picked up the pace in a half walk, half run, keeping my eyes on the creature until out of view, then I began to run.

When we arrived home, and for days after, we told everyone we saw. I don't think anyone really believed us. We went back the next day and found the tracks still in the dirt alongside the road. At that time, I didn't think to take a picture or to even measure the tracks, other than with our own feet. I estimated the tracks to have been eighteen inches long and about six inches wide. There were five toes and the foot shape was similar to a human, but a lot squarer.

It had to be very heavy, as the prints sank into the ground about a half-inch deep; ours didn't sink in at all from our weight, and the ground was hard. The tracks were visible for several days after when one friend, who half believed, asked me to go *[with him]* and take another look. Even after seeing the tracks, I still don't think he really believed.

Before this encounter, I had never really heard of bigfoot or gave it much thought. I had seen these kind of tracks in the past, while deer hunting deep in the woods, but I never really paid attention. I would look, wonder what made the track, and move on looking for deer. I was young, so it also never occurred to me that I would run into what made those tracks.

Note: When asked why they did not carry more ammunition with them that day, the answer was, "Hey, this is a reservation, we don't always have everything we need you know. I use to go hunting with one shell [one bullet]. That day I had a few shells, maybe a dozen or so, even if they were only .22 shells. The other guys were the same way; we use to take off with only a couple of shells because we did not always have enough to buy a lot to carry, plus the more you carried the less you can walk, so we always carried the least amount we needed and we did not figure on running into that thing that day!"

Scout Master, 1975

Submitted by: Clair L. Kuppenbender, age 76, of Canby, Oregon (retired; involved in many civic operations; current chairman of Canby Area Transit [C.A.T.]).

Encounter: January 1975, Lewis River near Battleground, Washington (Clark County).

An old friend of mine had become the scoutmaster of a local Boy Scout troop, and he invited me to join him as an assistant scoutmaster. Our troop was very active. We camped every month of the year and even owned our own old school bus for transportation. The kids had worked hard to buy this old bus and the local community college gave it a new paint job of orange and green, the colors the troop had chosen.

Our scheduled outing for January was for the weekend of the eleventh and twelfth, 1975. It was cold and rainy at home but we knew we would have shelters to sleep in. Those who signed up to go, about twenty boys, were what we called a Tenderfoot Troop. We had very few seasoned scouts. Most of the boys were eleven and twelve.

We boarded the bus early on Saturday morning and I followed with my car because we never took an outing without a car, just in case of an emergency. We were scheduled to camp at a Boy Scout camp belonging to the Columbia–Pacific Council. It was located on the Lewis River, several miles from Battleground, Washington.

This scout camp was directly across the river from a county park. There had been bigfoot sightings in the Battleground area and along the Lewis River. We arrived at the camp, drove down a muddy and rutted road about a mile to the parking lot. We loaded the kids up with their gear and started down into the river valley, a hike of about one and one-half miles. It was slippery with mud and snow and it seemed to be getting colder.

Upon arriving in the camp, we discussed staying. It was shortly after noon and we decided to wait awhile. In the meantime, the kids picked out their adirondack to sleep in (an adirondack is a three-sided building with double bunks on each end and along the inside. These had eight *[boys]* to a building).

As the day developed, things were going fine, so, by dusk, we decided to stay since the kids were all okay. These buildings were spread around and were perhaps 100 feet or so apart and open in different directions.

We fed the kids and got them into bed early. While Jim and I were having our last cup of coffee before retiring, we heard noises off in the distance. Strange and eerie howls, growls and whines, like nothing either he or I had ever heard before. We were both experienced with night noises in the woods, including bears, wolves, coyotes, elk, deer, etc. This was nothing like any of them. And we were not near any civilization.

We finally retired. Our unit had a step up into it of perhaps two feet. Jim took the lower bunk and I took the bunk above him. We were on the end of the building.

Sometime during the night, I awoke and heard a noise like someone moving around. I assumed Jim had gotten up to relieve himself. I sat up on the bunk and said something to Jim. I didn't realize it, but he was sound asleep below me.

It was an extremely dark night. No moon, no stars, and a heavy overcast—about as dark as being in a cave. I wasn't frightened in the true sense, but felt very hesitant. And did not have the guts to turn on a flashlight.

While seated, my eyes roughly eight to nine feet from the ground, I found myself looking directly into a pair of eyes! Face to face with some kind of creature. The hair on the back of my neck was standing up. The creature was at least eight to nine feet tall with very intense eyes. Suddenly there was movement and the creature quickly took off outside, and I could hear the brush loudly breaking as it quickly ran away into the woods.

The only brush around us was vine maple, a very springy wood that does not cut or break easily. I called to my buddy and woke him up. He saw nothing but heard the creature crashing through the brush and had a very queasy and uneasy feeling.

The next morning, we looked around and did find broken branches of vine maple (which even with a sharp ax are difficult to break), but no tracks. The ground was covered with wet maple leaves and the snow had melted, but no mud in that area. We fixed a good breakfast, broke camp early and never mentioned it to the kids until many months later.

Another time, Jim and I were up at Spirit Lake, on the slopes of Mt. St. Helens before she blew. We were having coffee with Harry Truman in his little café on the lake (he is the one who refused to leave the mountain). Two men came in who claimed to be professional bigfoot hunters. They claimed that the night before, they were tracking a bigfoot when suddenly he was tracking them.

I have no idea how truthful their story was. There had been sightings in the Mt. St. Helens area. I do know that the night at the scout camp, I was looking into the eyes of something standing on its two feet, roughly eight feet tall or so, and that is far too large for any bear in the area. The sounds we heard earlier were nothing we had ever experienced in the woods. We also never scheduled this campsite again.

Sasquatch on I-5 Freeway, 1975

Submitted by: Ross Peck, age 71, of Ferndale, Washington (retired school custodian).

Sighting: spring 1975, Interstate-5 freeway near Ferndale, Washington (Whatcom County).

It was a clear spring night with lots of moonlight, about two in the morning. I had worked late that night and was driving up Interstate-5 north of Ferndale, Washington, on my way home. Suddenly, about fifty to seventy-five feet in front of me, something leaped out of the brush on the side of the freeway.

It ran across my lanes (northbound), continued across the southbound lanes and disappeared into the trees and fields on the west side of the freeway.

It was a large, dark, hairy, man-like creature, at least seven to eight feet tall, running upright on two feet, with something light or white slung over its shoulders. I thought it was carrying a sheep or a goat.

It ran very fast and leaped the four-foot fence effortlessly. I slowed down instantly and thought, "Wow, what was that?" thinking it was pretty amazing.

I've told this story to friends and relatives and they usually smile and nod like, "yeah, uh-huh, sure." But I know what I saw. I don't know what else it could have been, other than a sasquatch.

It was much taller than the average height of a man, and taller than a bear, and pretty hefty looking. And running swiftly on two legs, carrying something over its shoulder.

Bathroom Break, 1975

Submitted by: Billie-Jo Ann Miller, age 42, of Cottage Grove, Oregon (award winning freelance photographer).

Encounter: summer 1975, Fort Lewis Army Reservation near Roy, Washington (Pierce County).

In the late summer of 1975, I was eleven years old. It was a warm, clear evening when my family and I were on our way home to Tacoma, Washington, from a camping trip. It was still light out, around 7:00 p.m., when we pulled onto a small dirt road on the forested Fort Lewis army reservation near Roy, Washington, to take a break from the ride. My seventeen-year-old sister and I walked a little ways into the woods to have a bathroom break, while the rest of my family were on the other side of our camper truck, poking sticks into a giant ant hill.

My sister and I had just finished doing what we had to do and started to walk the few yards back to the rest of the family, when we both had a funny feeling come over us. We both stopped and turned around. There he was, standing about fifteen feet away leaning against a tree watching us.

He was a dark, reddish-brown color, and covered entirely in hair except for parts of his face. He was about eight to nine feet tall, I am sure of this since his head was as high as the branches on the tree he was leaning against. Neither one of us said a word; we just looked at each other and began to walk real fast back to the front of the camper where we knew the camera was on the seat. I reached in, grabbed the camera and turned around, but he was gone.

We ran around to the other side of the camper and told my dad, who right away didn't believe us and wouldn't even look to see if there were any signs of him being there. He just told us to get in the camper and we left.

I will never forget the feeling of being watched, but never once did either of us feel threatened. My sister and I both felt that he actually wanted us to see him standing there, even though he made no noise at all.

Sighting at Asgard Pass, 1976

Submitted by: Michael Gordon Jameson, age 45, of Tacoma, Washington (silver and turquoise worker).

Sighting: July 1976, the Wenatchee Forest near Leavenworth, Washington (Chelan County).

I spent many summers camping and hiking around the highlands of the Cascade Mountain Range. In the summer of 1976, my friend and I hiked fifteen miles up several remote trails in the Mt. Stuart mountain range—where Asgard Pass meets with Dragon Tail, just below the Lost World Plateau, and where the Enchantment Lakes rest above the timberline.

We stopped to rest near a little outlet on the side of a hill where the water flows down out of the mountain. This hillside leads to the other side of the mountain. We had just removed our backpacks when we looked out across the outlet and spotted what we at first thought was a bear. As we watched it going up the mountainside, we quickly realized it was not a bear.

It was at least eight feet tall, covered in brown hair, and was standing up on two feet. Its posture was that of a human as it moved, using its hands as a human would when pulling one's self up. It finally reached up to a high point and pulled itself up and over the top. It then turned sideways, side stepped, and turned to look down at us. It had a look of what I would say was, "I'll stay out of your way, I just needed a drink of water."

It then turned and walked upright off into the woods. I have no doubt it was a bigfoot we were fortunate enough to witness that day at the base of Asgard Pass.

At that time, I knew people were hunting bigfoot so I did not report it. I did not want to invite the hunters who would have flocked to the area to track it. It has kept to itself all these many years without harming any of us. It's an intelligent being, which means it can think. Why hunt something that is so peaceful? Let it live in peace.

"Its Not Human," 1976

Submitted by: Michael Schwenk, age 52, of Oregon City, Oregon (sales manager for a concrete company).

Encounter: fall 1976, near Coeur d' Alene Lake, Idaho (Benewah County).

During the cold fall of 1976, my friend Jeff and I left from Portland, Oregon in my 1972 Porsche; we were headed to my brother's house in Harrison, Idaho, a little town around Coeur d' Alene Lake. After about eight hours of driving it started to snow very hard. We had to slow to a crawl as we drove off of the I-90 Interstate and onto Highway 97, the road that takes you around the lake to Harrison.

We were both tired and needed to stretch our legs and use the bathroom. It was around 3:00 a.m. We had not seen another car for miles, and back then there were no houses for miles around. The snow was piled up on the shoulders of the road, so I just stopped in the middle of the road, turned off the car and we both got out to pee and stretch.

It was snowing so hard, near blizzard conditions, that when we got back in the car we both had snow all over our heads and clothes. That caused the windows to start fogging. I started the car and turned the defroster on full blast. My lights were on and we were just sitting there waiting for the glass to clear.

I pulled out my cigarettes and lit one up, and as I did I saw something out of the driver's side window. I had turned on the dome light while we were talking; I quickly shut it off and wiped the remaining fog off of the window. I saw a shadow walking toward the car and coming straight for my door. I locked it and stuck my face close to the window, wondering if maybe it was a person that was lost or stranded. As it came closer, it looked like it was wearing a fur coat, like an arctic parka, clear up and around the face. I said to Jeff, "What the hell is that?" At that moment, this whatever it was pulled on my driver's door handle. When the door would not open, it calmly turned and walked around to the other side.

The rear window had a built-in defroster, so it was already clear from the fog. As it walked around the back, I could see its shadow in my red taillights. What I saw was freaky. It was huge and furry. If it had not been on two legs, I would have thought it was a bear. Jeff kept screaming, "Go! Go! Go!" I pushed the clutch in and tried to get out of there as fast as possible, but being snowy, the tires were just spinning. As I began to gain some forward movement, the thing had reached Jeff's door and tried to lift the handle. Jeff looked directly into its face and kept saying over and over, "It's not a person! It's not a person!"

Whatever or whoever it was scared the hell out of us. It appeared to come from out of the woods, and was covered in thick dark fur, which was covered with snow. It was very tall, at least seven feet, and very heavy. I never saw the face, but I did notice that it had very long arms and slouched as it walked with a weird pace. Jeff said he saw the face fairly close, but could not make it out as the face was obscured, as if bundled up in a hood with fur. More likely a lot of matted hair/fur covered with snow is what hid his face. I do not believe it was a man in a parka. Can I say for a fact that it was fur? No. Can I say with certainty that I believe it was fur? Yes.

Jeff and I have talked about this night over the years and to this day, we both agree that whatever it was, it was not a man. If it was someone stranded and in need of help, they'd have made an attempt to call out or bang on the window; not just pull on my door handle and when it would not open, walk behind to the other side. Another thing to consider is that both of us were six feet tall and 180 pounds. And this thing did not seem concerned with our size or the fact that we could have had a weapon. Given its size, it could have easily broken a window, yet it did not. Only IT knows why.

Dark and Stormy Night, 1976

Submitted by: M.M., age 37, of Kent, Washington.

Incident: fall 1976, Lummi Indian Reservation, near Ferndale, Washington (Whatcom County).

Once when I was a young boy, I was staying at my aunt's house for the weekend. Her home was surrounded by woods and located on the Lummi Indian Reservation. This particular night it was storming outside, with strong winds and loud thunder and lightning, when suddenly the power went out.

It was darker than dark. Everyone had finished their dinner and I was sitting at the table. I remember clearly, I was eating a piece of chicken when the lightning flashed and I saw a huge figure in the front window.

The curtains were closed, so I couldn't see exactly what it was, but I saw the outline of a very big, hairy-looking creature with no neck and a very wide frame as it moved past the window upright on two legs. It scared me so bad, the chicken fell out of my mouth and I got under the table. I didn't sleep much that night.

The next morning, I went outside with my cousins and found very large footprints flattened in the grass near the front of the window, which left no doubt that something had been out there.

Note: In the late summer and fall of 1975 and 1976, there were a series of bigfoot incidents that took place on the Lummi Reservation and around Whatcom County in general. These were experienced by myself (as related in Gram's story) as well as a few people I personally interviewed at that time. I recall several articles appeared in the local newspaper recounting some of the incidents.

Goat Attack, 1977

Submitted by: K.I.D., age 34, of Albany, Oregon (reporter and columnist for a pro wrestling website).

Incident: late spring 1977, mountain range in Douglas County, Oregon.

My father had bought thirty-five acres of woods situated on top of a small mountain range out by Camas Valley, Oregon, on which he began building a log cabin. One night around 2:00 a.m., he was awakened by the sound of breaking glass. He grabbed his gun and went out on the porch. He was confused; he had two attack-trained German shepherds, yet it was absolutely silent outside. Unusually quiet. He went back inside and grabbed a flashlight. The porch light was on, but it wasn't bright enough to see the animals.

As he walked by the car, he noticed broken glass on the ground. He raised the flashlight and saw that the back window of the car had been broken. As he peered into the car, he saw the reflection of animal eyes. He opened the back door and was stunned. On the floor, cowering as far under the seat as possible, were his two attack-trained dogs.

Apparently, something had frightened them so badly that they broke through the window to get inside the car. They whimpered and whined when he told them to "Get out." Finally, after repeated commands, my father reached in and physically dragged out both 100-pound dogs. They were shaking and very noticeably frightened.

About ten feet away from the car, my father noticed one of his goats lying on the ground. He had two medium - size goats, maybe thirty-five pounds each. They were each on a twelve-foot chain, staked to the ground. When my father touched the goat, he realized it was dead. He followed the other chain to the end and only found the other goat's head. His torso was missing. There was a lot of blood in the area the goat was killed and a small trail that traveled for about a hundred feet. By this time my father was pretty freaked out, so he took both dogs and went back into the house.

I'm not really sure what happened during the night, but when I got up early in the morning, there were three hunters there with their hunting dogs ready for a search. I wasn't allowed to go, but I was told what happened.

Apparently the dogs, which were trained to sniff out cougar and bear, never got one single scent. They went out searching until 4:00 p.m. with almost no results. The only trace of anything found was a rather large pile of scat, twelve inches in circumference and about six inches tall, with some white goat hair in it. The hunters and my father talked about it all night. That is the first time I can ever remember hearing the name bigfoot.

I was told the untouched goat had died of fright, literally. We do not believe it was a bear. I lived there for seventeen years; the only bear rarely sighted in that area were small black bear that never did any harm. Nor would they be enough to scare the dogs like that. Nor did the search dogs pick up any scent of a bear. I know from what I've heard and read about bigfoot, that this almost seems too bold and violent for him. But we could never figure out any other explanation.

Selah Sasquatch, 1977

Submitted by: Donna Headding, age 58, of Boardman, Oregon (forklift, heavy material equipment sales, and photography).

Sighting: Summer 1977,. Selah, Washington (Yakima County).

There is a gap between Selah and Yakima (kind of like a saddle between two mountains), and on the west side of this gap stands a hillside about two miles high, a place where ,back in the 1960s, I used to rent horses to ride for five dollars an hour.

There are old railroad tracks no longer used, and an old abandoned mine that goes down into the hillside. I don't know the history of the mine, but it is still there. Things change; highways and roads change, but the hillside and the old mine stay the same.

It was 11:00 a.m. on a nice, clear summer day. My husband and I were driving to Yakima and headed south on Highway 12. The long road travels through the gap for about two long blocks. At this point I looked off to my right, and there on the hillside, about five hundred yards away, stood a bigfoot!

He/she was standing motionless, was dark brown in color and covered in hair. I guessed his height to be about six feet (possibly a junior bigfoot). His arms were at his sides; I could see the gap between his arms and his body. He stood very still as not to be seen, but he was very clear to me.

I got very excited and began making a fuss, "See him? See him?" My husband looked up to see him as well. He had to pay attention to the road and keep driving, as traffic was busy. There was no shoulder to pull over or we would have stopped to better check it out. I watched it excitedly until I couldn't see it anymore.

I grew up hunting and fishing, and with my history of being familiar with this area, I know what animals and critters are common there. You never see any animals up on the ridge. Nor was it a person in some sort of monkey suit; no one would dress up in a suit and stand clear up on that hill. I know exactly what it was.

I thought later about how he was standing up where there was no path, just looking down the hillside. He seemed to be figuring out how he was going to cross the Yakima River, which was about 400 feet in front of him, and then the busy four-lane highway. Why else would he put himself into that predicament by standing out in the open (in the gap) and in broad daylight?

The way I think he was headed was toward the Firing Center Land, a military training center that is off limits to the public all the way to the Columbia River. After swimming the Columbia he would have nothing but farmland to cross clear to the Blue Mountains.

We have told many people of our sighting over the years, and most don't know what to think. Having all these years to think about it, I know now that we did the right thing by not stopping, not pointing him out to others, and letting him be. His sight is haunting. I saw something very special that day; I consider it a gift. Truly an experience I will never forget.

Blue Eyes, 1979

Submitted by: D.J.G., age 41, of Vancouver, Washington (technical writer).

Encounter: August 1979, along the Lewis River in Skamania County, Washington.

The incident I want to tell you about happened when I was a young boy, and my mother was dating my soon-to-be stepfather, Ed. One nice summer day, mom, Ed, my little brother, and I all went on a picnic up along the Lewis River in Skamania County. Ed used to pan for gold and knew all of these remote spots. It was a picnic, at least for the two lovebirds.

This area is so remote that no other people or cars were seen the entire time that we were there. Looking around, I spotted an old set of railroad tracks that for sure were not being used anymore. My little brother and I asked for permission to go exploring. We were obliged, as the grown-ups had other ideas.

The two of us followed the tracks for a ways when we came upon a long-inactive train tunnel. We wandered inside the old, cool, dark, and moist tunnel that had moss on the walls and had a slight curve to it. We walked for a while, when suddenly the hair stood up on the back of my neck, I began to get an eerie feeling that we were being watched and sensed by an animal.

As we continued, we passed the curve in the tunnel enough to see the light at the other end. The end was far enough away that the opening looked about the size of a dinner plate. I could see something standing in the middle of the opening, but wasn't sure if it was a large branch hanging down or something else. As we got closer, to maybe 600 feet, I could see that it was not something hanging down because there was a gap over its head. I could now see it was some kind of animal that was standing straight up. My first thought was that it was a bear. I knew how they stood up on hind legs and almost looked human like.

We continued to walk toward it, half out of fear that if we turned to run it might cause the animal to have a chase response, and the other half out of the foolish fearlessness of two children ages nine and twelve. The closer we got to the thing it became clear that it was not a bear. The first thing that tipped me off was how long the arms were. They just hung at the creature's sides like two long poles. Then it moved, just slightly, and I could see that it was so tall that it reached almost to the top of the train tunnel.

I looked behind us; I could now not see the other opening. We had walked far enough along the bend of the tunnel that there was no light visible to our rear. I wondered, did that mean that the animal could not see us? How good was its vision? The creature then spread its arms out so they were reaching diagonal. It was blocking the tunnel exit. As if it did not want us to pass.

I took my little brother's hand as we stood there in shock. Our breathing was in sync. We moved deliberately closer to the creature, coming within 100 feet of him. To my surprise and utter and complete horror, there stood this beast of proportions almost unimaginable.

This thing was beyond beast. It was huge, enormous, the granddad of all bigfoot! It seemed like it filled up the whole tunnel. Its entire muscular body was covered in long brown hair, which was sticking out in each and every direction. The hair looked bristly, as though every muscle on the surface of its body was flexed and the energy from its body was crackling through its hair. Its head was much larger than pictures I have seen, and it had a neck. It had monstrous calves and I have no doubt it could have outrun a horse. This thing was like no zoo animal I had ever seen. This thing was bigger than I've heard bigfoot are suppose to get. He must have been the "André the Giant" of bigfoot.

Then, suddenly something happened. Both my brother and I instinctively froze as the creature stepped back a few strides from the entrance of the tunnel, thereby showing himself clearly in the sunlight. We moved closer, coming within twenty feet of him. I could now see his eyes, which were a light and brilliant blue. And they glistened with intelligence. His hair sparkled a golden brown in the sunlight taking on a satiny sheen.

He was not afraid and wanted us both to see him in the sunlight. He was the most magnificent creature I had ever seen. I could also see that he was old. He had an ancient presence or wisdom about him—so aware and conscious. There was no dumfounded look in his eyes, like there might be with an ape, which I kept thinking he must be.

He had a smell about him that reminded me of several things. Like the feed corn that they pile up for cows, the smell of fresh moist soil, and the smell of the paste that they give you in the first grade to glue your art projects with. At first it was not pleasant. Then, it seemed to be kind of welcoming. It evoked in me a feeling of love, like for a large fluffy dog that was wet or damp, yet there was a certain freshness to it as well.

He clearly let us know that he liked us and took great pleasure in letting us see him. Almost like a wise old grandpa showing you something that he knows will shock, surprise, and please you, and that he knows will change you for the good.

He seemed like he wanted to take us in his arms and hold us there for awhile. Just squeeze us and let us know just how much affection he had for our species. As if a lifetime of love and curiosity for humans had grown in him, and now in his presence were two small human creatures that he adored.

It seemed we spent a lifetime with him. Like we shared time in his world and learned so much from him. I felt wiser afterward, and more loving. Less restless for my age.

At one point, his eyes teared. That's when I knew it was time to go. He had arrived at the point emotionally where he sort of woke up and remembered that he was he, and we were we, and there was no way that we could be friends. At that point I could imagine that bigfoot were truly standing near the campfires of the Indians, as legends of late told. That they had got that close, yet knew, in order to keep that necessary separation between bigfoot and human intact, they could never get too close.

My nose stung and though I tried to fight it, my mouth pulled down on both sides as if to cry. I knew that I would never see him again.

Clutching my brother's hand tightly and without taking our eyes from his, we began to walk backwards, back toward the other end of the tunnel. We walked backwards all the way to the bend in the tunnel and never once did we slip on the railroad cross ties. Our adrenaline so high that our senses did not need to see.

By the time we got back to the picnic, they were packing things up. Evidently, red ants had invited themselves to lunch and we were leaving. My brother and I climbed into the back of the little red Mazda pickup and simply let the wind blow over us for the next fifty miles home. We have never spoken of this, ever.

It has been twenty-eight years now, and this is the first time that I have shared my story. I feel relieved to be getting it off my chest. However, I need no one to validate what I saw.

The Hunting Trip, 1979

Submitted by: Brian Glaholt, age 39, of Des Moines, Washington (garage door installer).

Encounter: summer 1979, forested area between Mt. Rainier and Mt. St. Helens, Washington State.

When I was thirteen years old, I went deer hunting with my dad and his friend Rodney. We headed up to a forest somewhere between Mt. Rainier and Mt. St. Helens. When we arrived at the trailhead it was still light out, so we started up the hill. By the time we had hiked half way up, it began to get dark and we couldn't see very well.

We had found an old miner's cabin earlier and decided to hike back and camp there until morning, rather than hike all the way back down the trail to the truck and camper. The little shack of a cabin was very small, and the only thing in it was an old wood stove. We bedded down and soon went to sleep.

Sometime later, I woke up to my dad telling me to "Quit kicking the door!" I said to him, "I'm not kicking the door." I fell back asleep only to have my dad again yell, "Quit kicking the door!"

This time I sat up and said, "Dad, I'm not kicking the damn door!" I think he was shocked by me using the word "damn" and it finally sunk in that it wasn't me at the door. He told me to be quiet and he woke Rodney. The three of us sat there watching as the door of the cabin shook violently.

We were all very well loaded with fire power; I had a

30:30 rifle, my dad had his 30:06, as did Rodney, as well as his .44 magnum, which he said was for the bears, just in case. The next thing we knew, whatever it was began pounding on the low roof of the cabin—boom! boom! boom!—which was followed by loud grunting sounds and a very fowl, disgusting smell. I looked over at my dad; when you're thirteen years old and you see your dad's eyes pop out of their sockets, you get real scared, real fast.

We got our guns, ready to blast whatever it was to pieces when my dad said, "Hold! Don't fire yet." He then aimed and shot one round through the ceiling, which was followed by a high-pitched scream, and then whatever it was took off through the nearby bushes. My dad turned to Rodney and said, "That thing ran away on two feet!"

We waited a while and, not hearing it return, decided to head back down the trail to the truck. It was very dark outside and I was really scared, so they gave me the flashlight and said, "Go!"

Well, I bolted down that trail so fast, nothing was going to catch me, not even my dad or Rodney, whom I left back there with no light! Hoping my old man would understand my running away at full speed, I went back and found them, got yelled at, and then we all started down the hill together.

When we got to the truck, my dad and I crawled inside the back camper, while Rodney slept in the cab of the truck. Before too long, something began violently rocking the truck back and forth. Good old Rodney immediately started up the truck and took off flying down the gravel road and away from whatever it was that was scaring the hell out of us.

We decided we had enough of hunting and would go home. The problem was that in order to leave, we had to go back the way we had just fled. We piled into the front of the truck, turned it around and started back. When we were almost to the spot where the truck had been parked we came upon the creature.

We got a good look as we swerved around it. It was

big, real big, probably well over seven feet tall, and a good 550 to 600 pounds. It was covered in dark, brownish hair and it appeared to have no neck. It was walking on two feet with its huge arms swaying back and forth way out in front of its body. We darted past the creature and quickly left the area, never to return.

Not one of us has ever gone hunting since. I do know that it was something I will never, ever forget. This is a true story.

Baby Bigfoot, 1980

Submitted by: D.A.F., age 39, of Eugene, Oregon (heavy equipment rental manager).

Encounter: spring 1980, south hills of Eugene, Oregon (Lane County).

When I was twelve years old, my fourteen-year-old cousin and I were dirt-bike riding in the south hills of Eugene, Oregon. Here are thick-forested areas that reach hundreds of miles from the mountains all the way to the coast of Oregon. My cousin lived next to the forest where fire trails were cut throughout and we found these perfect for dirt bikes.

We were having a blast as we rode through the dirt trails weaving up and down and flying off the jumps. While riding past one thick patch of brush and trees, we smelled a strong and foul odor, like something dead. And suddenly, standing in front of us along the edge of the tree line, stood a small creature about four feet tall. It was not a dog, not a deer, not a bear; it was not like anything we had ever seen before, and it stood upright on two feet.

He was standing just twenty feet away from us. He had thick, dark brown hair, about six to eight inches long, that hung on him like dreadlocks. He had big eyes. He just stood there in one spot staring wide-eyed at us, as we did him, for maybe two whole minutes. He stood perfectly still, as if he thought that maybe we wouldn't see him standing in that tree line. Suddenly a loud grunt came from behind the creature, and it turned and ran back into the thick woods. We wanted to chase after it, but the woods were too thick for a dirt bike. And we weren't sure we should go after it anyway.

Our grandpa was a park and recreation director for the Lane County Parks Department. He used to tell us stories about bigfoot in Washington, so it didn't take us long to figure out what we had encountered. I am in the woods all the time, camping, fishing and four wheeling. I have never seen or smelled anything like it since, but if I smell it again I will know what it is.

A Tale of Two Encounters, 1980

Submitted by: Charlie Darnay, age 52, of Wasilla, Alaska (U.S. Army Infantry ten years; webmaster, copywriter, IT Technician).

First encounter: May 1980, in the Three Sisters Mountain area, Oregon (Deschutes County).

In the late spring of 1980, I was traveling with my family from California to Washington State. My wife and I were in one truck, and my father, mother, and uncle were in another truck behind. I had just pulled into a campground just below the Three Sisters Mountains area of Oregon, and it was very late at night. I'd been driving for many hours and the coffee that had been keeping me awake now was demanding to be released. I stepped out of the truck and walked a few feet into the dark woods, lit only by faint moonlight and the truck headlights reflecting off the trees.

I had just started to answer nature's call when I felt a presence. As soon as this feeling came to me, I was assaulted by a smell—musky and oppressive. I turned, and in the dim light behind me, where I expected to perhaps spot a skunk , I saw a hairy chest about twenty feet away. I looked up and large brown eyes met mine. Before I could react, the creature bolted almost silently into the woods.

It was close enough that I could tell it was entirely covered in hair, was at least three feet taller than me (I'm five foot, eleven inches) and when it turned and slipped away, it did not drop to all fours like a bear would have. I have encountered several black bear and they always react the same, running away on all fours. This creature just stepped off leading with its right foot and was gone in seconds.

I was still zipping up as I landed in the truck seat, locked the doors and grabbed the CB mike, all in one fluid motion. I told my dad, several minutes behind us in another rig, that I had just seen a bigfoot behind the truck and I'd be waiting for him at the park entrance. I was still shaking and almost couldn't talk when they rolled up. After excitedly telling them of the encounter, I began calming down and became angry with myself for reacting the way I did. We went back to the same site, spent the night and I stayed up until almost sunrise hoping to glimpse the creature again. It is amazing how many sounds the woods make at night when you're sharply attuned to them.

Other than the terror, the encounter felt like an encounter with a sentient being. The same feeling you get when face to face with a mountain gorilla or a beluga whale at the zoo.

Second incident: spring 1981, near Republic, Washington (Ferry County).

About one year later we were living in the Okanogan Valley. I was attending school in Spokane and decided to drive home for the weekend. I was just outside the small town of Republic, in Ferry County, when my car broke down. I got it off the road and started walking. It was after 2:00 a.m., and I was about twenty-five miles away from my dad's house, so I expected I might actually end up walking the entire distance. The bars were already closed and dad didn't have a phone line up to the house yet.

I made my way through Republic and was starting down the long grade west of town. The road went through a very heavily wooded area, no lights other than from the crescent moon. Suddenly, I heard a noise in the bushes across the road. I listened intently as I walked along and decided that whatever it was, it was much bigger then me and sounded like it was walking on two legs.

I could hear the crunch of branches and the brushing of trees. I had decided that it was either a cougar or a bear, both being plentiful in the area. I knew that cougar might stalk like this thing seemed to be doing, but I didn't think a black bear would stalk a man. As I rounded the turn in the road, the wind shifted and I caught the hint of a familiar smell.

Knowing I had nowhere to run and nothing but roadside rocks to defend myself, I continued my walk, pretending not to notice I was being tracked from the shadows. There was a little moonlight, but try as I might, I could not see into the trees across the road. By the time I reached the bottom of the hill, the smell had grown stronger and unmistakable as the same smell I encountered in Oregon almost a year before.

I didn't manage to muster the nerve to cross the road and confront my tracker, and after about twenty minutes of watching and pacing me from the shadows, it left me alone. Another twenty minutes and my heart finally returned to its normal rhythm.

Note: The Three Sisters are three volcanic mountains that sit in the Cascade Range, Oregon. They were named Faith, Hope, and Charity by early settlers. The surrounding wilderness, which contains a total of 286,708 acres, is managed by the U.S. Forest Service.

Mystery at Larch Mountain, 1982

Submitted by: M.T.P., age 43, of Vancouver, Washington.

Sighting: February 1982, Larch Mountain in the Cascades near Mt. Hood, Oregon (Clark County).

My name is Mike, and when I was nineteen, my friend (also named Mike) and I had an experience we never will forget. We got up early that morning in February to head up the historic Columbia River Highway to Larch Mountain in the Cascades near Mt. Hood. We were going up to put my Ford Pinto to the test in the snow and to check out the summit lookout on the mountain.

The lookout is a fenced-in, oval shaped area about the size of a normal bedroom. From this lookout you can see five snowcapped volcanoes, the Columbia River Gorge, and on a clear day, four more volcanoes and the city of Portland. It is a spectacular view.

We arrived in the empty parking lot of the lookout about eight or nine o'clock in the morning. We both glanced up to the fenced-off lookout area at the same time and noticed something brown, hairy, and large standing up there. This thing was huge and appeared to take up half the size of the lookout area. From our distance of about 100 yards, I would estimate this thing was at least four feet wide, had no neck and a huge head.

I knew it wasn't human. Mike and I both said at the same time, "bigfoot." It looked at us for about ten seconds, turned to its right, bent down and picked up what looked like a large cardboard box, its left arm fully extended to hold this thing. It stood up and walked off with the other arm swinging.

We immediately started for the lookout in a half run-half walk; we got there and saw nothing. We began to look out over the fence in all directions, talking to each other about where it might be and what direction it took. And here is where the story gets stranger.

There is one other trail (at least a five-mile hike) that concludes at the lookout through a gate about four feet wide. Not more than two minutes had passed when two guys show up walking through this gate side by side (I think human behavior is to lead or follow). They never took in the view, in fact they moved to the side opposite of the view.

Now get this, these guys looked just like two Richie

Cunninghams taken right off the set of *Happy Days*. It was the middle of winter in the Cascades and these two guys are dressed in short sleeved, fifties-era shirts, which were tucked neatly into tan dress slacks, brown belts, and brown dress shoes. While I am taking all this in, Mike asked them, "Did you see anybody up here?"

Without looking at us, they both responded (at the same time), "A woman in a fur coat." Then they immediately turned and walked back down the trail from which they came.

Okay, here is my argument why this can't be. We were the only car in the parking lot. It was very cold; we were wearing hats and coats. The only other trail to the lookout begins at Multnomah Falls in the Columbia Gorge, which is about a five-mile hike to the lookout.

So, here are two guys with no coats or winter-wear neatly dressed with not a hair out of place after just hiking five miles. Another thing is, you would think after hiking that far to the lookout they'd have stayed more than a couple minutes, and would've checked out the view.

The next is—a lady in a fir coat? And all alone on a five-mile hike in the winter time? Also, when we originally saw the creature, at the fence line along the lookout, the top rail came up to about the groin area of this creature. When I reached the same spot, I noticed the fence's top rail was just below my shoulders. Now I'm five foot eight and I've never seen a lady eight feet tall, have you?

Given the years we have had to discuss and think about things, Mike and I feel that there is a little more to bigfoot. A cover up of sightings? Extraterrestrials? I have only told this story to a few family members and close friends; I have been waiting for an opportunity to share on a larger scale.

Campsite Tracks, 1983

Submitted by: R.J.C., age 56, of Everson, Washington (tool room tech., college graduate, Nooksack Indian).

Incident: fall of 1983. Grouse Butte, Mt. Baker National Forest, Washington State (Whatcom County).

This story begins in the fall of 1983 on a camping trip to Grouse Butte (take Glacier Creek Road up to the very end of the road, west of Mt. Baker). My youngest son and I went out hiking, intending to follow the gravel road back to camp. As we approached the road on the far side of the mountain, we saw our Chevy approach with my wife and oldest son all packed and ready to go home. The oldest son, then twelve years old, had found large, bare footprints that had circled our tent. He then showed the prints to mom, which created an urgent need to get out of the woods. That was her last camping trip.

If it was a hoax, it was a very good one. The ground was cold and extremely firm. We did not even leave a print. Whatever left the prints was not just large, judging by the size of the eighteen-inch long foot print, but heavy as well. Those big foot prints sank a good inch into the near frozen dirt. My eldest son has been a die-hard believer in bigfoot to this day.

Son's Account of Tracks

R.D.C. Jr., age 36, of Everson, Washington (construction field).

My parents used to be avid campers/hikers. They had brought my brothers and me on several camping trips to local wilderness areas. This particular trip was in the winter. We had driven several logging roads looking for an area to camp; finally we came across a spot that was near Glacier, Washington.

On the first day, we set up camp at the end of a logging road, basically in the middle of the road, near a large area that had been clear-cut. Our camp was surrounded by logging slash as well as live shrubs, and the weather that day was cold, cloudy and the ground was frosted, almost frozen. During the afternoon the ground was semi-ridged, but in the morning/evening hours the ground was almost as hard as concrete.

In morning, I woke up early. My father and one of my brothers had already gone hiking. I had to do the morning business so I walked away from the tent and stood near some shrubs. As I finished my business, I looked down at the ground and noticed tracks. Because these tracks were in grub area they were hard to make out, but it seemed that the tracks were headed toward our camp.

I followed the tracks and when I came to the road, I found them to be clearly imprinted into the ground. I knew when I saw the first ones that they had to be bigfoot prints since they resembled human footprints. I would estimate them to be sixteen to eighteen inches long, and there were five distinct toes. Imagine human prints in moist sand, that is exactly how they looked but in frozen ground, and much larger, something no human could do.

I followed the tracks right to our tent! I yelled for my mother, "Mom! You need to come out here!" By this time I had traversed the tent in a complete circle and was now following the tracks back toward the grub. The stride of the tracks on the road were approximately thirty to thirty-two inches apart, the stride around the tent was closer together, maybe twenty-six to thirty inches apart.

My mother came out of the tent to ask why I was yelling and I told her to come look at the tracks. Now, we were used to seeing raccoon tracks (which could not imprint frozen ground), deer tracks (maybe scuff the ground), bear tracks (but never near our camp). I showed her the tracks at the edge of the road and told her that they go all the way around our tent. We remembered hearing rustling sounds during the night, but they didn't seem to be this close to camp.

My mother is not one to scare easily, but she immediately went back to the tent and started packing up our gear. Once we were packed up and the Blazer loaded, we moved from the camp area to an intersection to wait for my father and brother to return.

You know, being Native Americans we all grew up with the stories our elders passed on. I think my mother was not a believer until after that day. It really creeped her out and I think she is pretty rugged as far as creatures go. She used to scuba dive, hike, bike, and canoe, so we spent a lot of time out in the wilds and she did not scare easily.

Creature on the Hill, 1984

Submitted by: Robert DeWitt Jr., age 37, of Tacoma, Washington (maintenance foreman).

Sighting: fall 1984, Tahoma State Forest, Oregon (Lewis County).

When I was a child of about thirteen, I went deer hunting with my father and his friends to the Tahoma State Forest, west of Mt. Rainier. I was with my father walking a road under a very large hill, almost a mountain really. It was clear-cut from top to bottom for the most part. We were watching a group of our friends who were standing on a large rock outcropping; they were waving to us and we were waving back.

It was a clear day, and as I watched them, I noticed something moving off to their right at the same level of the hill, about 1,000 yards away. It was a large black creature walking on two legs across the stump field, heading straight toward them.

What was odd was the thing moved half way across the field in five minutes. I've walked across this part of the hill before and it took me at least thirty minutes, the stumps and crap were a nasty mess at the time.

Whatever it was walked upright and stopped about three times, once to reach down to seemingly play with something on the ground. It was much thicker in body than our friends, and a good foot or more taller than one friend, Harvey, who is well over six-foot. I watched it for a good ten minutes until I lost sight of it as it went into some alder saplings under and around the rock outcropping our friends were standing on.

For the most part I think I'm level headed and doubt some of the stories I've heard from people about bigfoot.

But I look back on it now in wonder. There was something up there that day!

So, I tell this story to my nieces and nephews and to those who bring it up. And before I leave this earth I would hope to meet another of their kind a little closer.

Creepy Camping, 1984

Submitted by: Michael C. Scott, age 40, of Sumner, Washington (warehouse manager).

First incident: August 1984, Republic, Washington (Ferry County).

In the month of August 1984, my dad and I took a road trip down to California and Nevada and then back up through Eastern Washington State. My story takes place somewhere around Republic, Washington, directly north of the Colville Indian Reservation in a mountainous region of Ferry County. All I know for sure is it was very dark, and I've never been so frightened in all my life—besides getting married.

I was sixteen years old and it was an exciting time, as I had just got my driver's license and Dad let me drive a little on our trip. We had our 1978 Ford pickup truck with a canopy on the back. I found out early on there was no way I could sleep back there with my dad because he snores so loud, so I began sleeping in the cab for most of the trip.

One night, we were having a hard time finding a place to park for sleep. We weren't hungry, just tired, so any place would do. We found ourselves high in the thick woods of the mountains, and like I said, it was dark, totally pitch black. No moonlight whatsoever.

My father found an old horse camp off to the side of the road, so we pulled back a hundred yards or so off the road and parked. There was one other camper about fifty yards away. My dad got right out of the cab and into the canopy to lay down. I told him I was going to sit up for a few, and I wondered over to a wooden picnic table. I put on my headphones and began to "adjust" my mental state.

As I sat there, an overwhelming fear suddenly came over me. I began looking around but it was so dark; I could not see more than seven to eight feet in front of me, and I couldn't hear because I had the Walkman turned up as loud as it could go. The fear became more intense with each passing second. I finally took my headphones off so I could get a better grasp on my surroundings. The silence was eerie.

It was too quiet. I felt I was not alone and had this feeling I was being watched. I tried to shake it off and decided to just go to bed and put myself out of my fearful state. I remember walking over to the truck while looking over each shoulder. I could hear my dad's snores and knew he was deep asleep. I opened the passenger side door of the truck and the dome light came on. Now, I could not see more than three feet around me with the light. With the door open, I began to remove my shoes when something told me to just get in the cab and close the door. I finished undressing in the cab and climbed into my sleeping bag.

At sixteen I was a tall kid. With my feet on the steering wheel side and my head on the passenger side, the top of my head touched the armrest on the passenger door, and if I wiggled my toes they touched the driver side door. No room to spare. I pulled the sleeping bag all the way up to my nose so I wouldn't get cold in the middle of the night.

Only two minutes passed since closing the truck door and getting into my sleeping bag when I begin to feel the truck rocking back and forth from what I thought was my dad rolling around. Suddenly, I realized my father was still snoring and in a deep sleep, and snoring people don't move while they are snoring! I lay there motionless when suddenly I heard the snap of twigs outside the truck. I felt the truck's motion from side to side intensify and then, just on the other side of the passenger side door, six inches from my head, I heard something rubbing up against the truck.

I told myself to look, it could be something cool like a

bear or an elk, or a bigfoot! I decided that if I lowered my sleeping bag and raised my head to the passenger-side window and bigfoot was on the other side looking back, I would actually die of fright! So I laid there for what seemed an eternity until the rocking and rubbing stopped. I eventually fell asleep without ever moving or looking.

The next morning I woke to my dad telling me what a beautiful morning it was and a cheerful, "How did you sleep?" I began to tell him everything (this farm boy from Minnesota who myself and my brothers would jump from anywhere possible to try and startle, but always to no avail). He laughed at me, slapped his thigh, and shook his head.

Not to be laughed at for long, I began looking around the picnic table, behind the truck and all around it. And then I saw it! Proof that something was just outside my door the previous night. Thick, black, coarse hair caught in the side mirror brackets that attached to the door. It actually happened, but what was it?

Second Incident: August 2004, Blewett Pass, Washington (Chelan County).

Twenty years later, again in the month of August, my girlfriend and I were doing some camping before flying off to Vegas for my birthday. We left Auburn right after work and headed over I-90 and then up over Blewett Pass trying to make it to Wenatchee. But as I was up late the night before and after working all day, I was very tired, so we began looking for spots just off the side of Blewett Pass for the night.

After a few non-campable stops down different dirt roads, we turned down the Old Blewett Highway. On the north side of the pass, we found a gold panning claim site off the side of the deserted road. I set up the tent, had a beer, took a couple of Tylenol PMs and soon fell fast asleep.

When I woke in the morning, my girlfriend told me that in the middle of the night, she was awakened by heavy footsteps and something moving around our small, two-man tent. It was heavy enough to break the thick twigs on the ground as it encircled the tent. Eventually, it stopped right above our heads hovering over us. She could see its shadow as it stood there, the moon casting its silhouette across the tent.

It had form like a man, yet it was huge, a very tall figure with huge arms. She says she nudged and kicked me, but due to my sleep-induced state, I thankfully did not wake up, leaving her scared to death until it finally moved away. It was dry and hot that summer and so the ground was too hard and dry to make out any actual prints of any kind. All the truth!

Back Country Bigfoot, 1985

Submitted by: Mary Long, age 42, of Denver, Colorado (public land surveyor – instrument operator)

Sighting: spring 1985, Trinity Alps, Northern California (Siskiyou County).

I have had two encounters in my life that I would call "bigfoot experiences." The first was after I joined the C.C.C. (California Conservation Corps). I chose to work in Montague California, Siskiyou County. We did a lot of work for the forest service, clearing trails, putting up barbwire fences in cattle ranges, etc.

One day in April of 1985 (it was a cool, brisk morning and the sun was out) I was sent out with a crew to do some trail work in the Trinity Alps wilderness and stayed in forest service barracks for the week. This project was located outside the small town of Callahan. When we arrived at the project, I was asked to do a job and stayed at the vehicle with another crew member, Rick. When we finished, we hiked down the trail to wait for the crew to come in for lunch. But, they never came back for lunch.

While we waited for the supervisor or someone else to show up, both Rick and I saw something step out of the tree line, thinking it could be someone coming to get us. Then we realized it wasn't some*body*, it was some*thing!* Something big, huge, extra large. This thing was about seven to eight feet in height, and as soon as we saw it, it whisked itself back into the tree line. But not before we got a good look at it.

The coloring of this creature was a blondish tan, and it stood on two legs. And it had a strong musky odor. It wasn't a bear, it wasn't a deer, and it wasn't human. Rick and I looked at each other wide-eyed, "We just saw a bigfoot!"

The second experience was a few weeks later in late May of 1985, when I was chosen to be on the Klamath–Sequoia back country trail crew. Our first hitch was outside of Seiad Valley; we had to hike up nine miles to get to camp. We were there from late May to maybe late July, reconstructing trails and rock walls. After getting to know the crew, you have your little "groups." The crew split up into two different crews to do two different projects, with my group staying at base camp.

One night, after some of the crew went to bed, there were four of us sitting around the campfire when we heard something in the brush, snorting, like maybe a horse would do. And we could feel the presence of something watching us. Although I was raised on a farm, I have never heard this kind of snort. In fact, "grunt" might describe it better.

We didn't know what to do, so we all escorted each other to our tents, grabbed our sleeping bags, and all slept in the kitchen tent. There we lay with our heads covered and didn't get much sleep. This snorting/grunt sound happened four different times in four nights, always around midnight. The forest service employee tried telling us that there were wild horses in the area, and this is what it could be. To this day, I know it wasn't wild horses. There were no hoof prints, no hoof stomping as if it was going to charge at us; it just made a sound that I have never heard before or since.

I have lived up against the Marble Mountain Wilderness for nine years of my life, and I've seen bears come and go. I've had a bear go through my front door to eat the cat and dog food, and even then I have never experienced what I did when I worked in the Trinity Alps Wilderness, or in the Klamath National Forest.

Kirkland Sighting, 1987

Submitted by: Cindy Baxter, age 43, of Kirkland, Washington (former junior high school teacher; now stay-at-home mom and home schooler of two).

Sighting: summer 1987, vicinity of Lake Washington High School, Kirkland, Washington (King County).

> My then-fiancé and I had a sighting while sitting and talking in front of my house late one summer night, across the street from Lake Washington high school. We are fully convinced we saw a bigfoot walk across the street about three houses down from us. The creature stepped out from a green belt that comes up from Lake Washington, took maybe five strides/steps and he was across the road, disappearing into a wooded area that continues east. We couldn't speak for several minutes but we both knew each other had seen it.
>
> We knew he was about eight feet tall because of where his head passed by on the metal reflector band of the power pole, which I had just been down by that morning to walk the dog. The dog had paused to go and I looked up, way up, and noticed the metal band on the power pole … Later that night, I realized how tall he had to be. He was covered in black hair and didn't seem to have much of a neck. He reminded me of what a football player with his gear on looks like in the neck.
>
> We were not going to say anything until we spotted an article in the newspaper about another bigfoot sighting in Eastside Kirkland. We reported our sighting to our friend

Kevin, who had been doing lectures and presentations of his footprints casts and other compiled evidence. Kevin came out and investigated the sight about a week later.

Kevin reenacted the scenario for us and walked the way most say bigfoot walks, hunched over a little, arms swinging. But Kevin (a former football player) could not make it across the street in five strides. It took him more like ten. It had been a week and no prints could be found by then, nor was there any evidence we could see. But we are convinced of what we saw.

Even though this was a frightening experience, it was also something that made me feel very alive afterward. We actually knew Kevin from college and in the past even joked a little about his bigfoot hobby. No more!!! We believe!

Memorial Day Mystery, 1988

Submitted by: R. P. White, age 49, of Fortuna, California (retired truck driver).

Sighting: Memorial Day weekend, 1988; Porcupine Lake in the Trinity Wilderness of Northern California.

My sighting occurred on Memorial Day weekend of 1988. My best friend, along with his dog, and I went fishing up at Porcupine Lake in the Trinity Wilderness. Once there, we found a dirt road on the map and drove for about six miles and parked the Jeep at the end of the road. We gathered our gear and hiked up hill for almost an hour. When we reached the top, we were on the rim of a crater and started our decent down the other side. It took another half-hour to get to the lake. Total time was approximately an hour and a half. We set up camp on a very large boulder overlooking the water and we fished the rest of the day.

We got up very early the next morning to fish; we had a great day catching German brown trout until we had about a dozen each. We packed up our gear late that afternoon and headed out. It took about forty-five minutes to hike back up the hill.

As we neared the top I noticed my friend's golden retriever was acting very peculiar. He was whining, and trying to hide behind his master's legs. This was not normal for an older dog with a lot of outdoor experience.

The next thing I noticed was a smell. It was the most awful thing I have ever smelled. I would describe it as a cross between fecal matter, perspiration, urine, and decayed flesh, and to top it off, skunk. I guess if you put everything you've ever smelled that was bad into one smell, and then multiplied it a hundred times, you would be close. I now think bigfoot uses this as a hypnotic tool.

I looked up in the same direction the dog was looking, to the left and about fifty feet behind us. That's when I saw her. She was standing in the manzanita. Her shoulders were six inches above the top of the brush. I estimate her to be a little over seven feet tall. Her coat was mostly dark; her face was a pale tan or light brown. She had gray, almost silver hair around her face. She had very dark brown pupils; they were larger than a human.

We made eye contact for about five minutes. It seemed like an hour, or more! It felt like we spoke in our minds. The first thing I recall was her curiosity, she was confused as to why the paralyzing smell didn't affect me the same way it was affecting my friend. And then I sensed she was telling me not to be afraid.

The next thing I did was look back to see what my friend was doing. He was standing there with a dumfounded look on his face. He was just staring blankly at nothing. It was like he was hypnotized! I elbowed him in the ribs. It had no effect, so I did it again. He was still dazed. The third time I elbowed him, I did it much harder and he snapped out of it with a start. I didn't have to point her out to him; he looked over at her and immediately pulled out his .357 magnum and pointed it at her. She quickly turned and retreated, in a zig-zag pattern. Just like a defensive maneuver used by the military, when you are being shot at. She was parting the manzanita in ten to twelve-foot bursts straight up the hill.

I don't recall getting the rest of the way up the hill or the decent down to the Jeep. The next memory I have is of being at the Jeep, and when I checked my watch I found it had taken almost four hours to get there. As you may remember it took an hour and a half to get to the lake. I have no explanation for the missing time.

We loaded the Jeep and drove back to Weaverville, California. We didn't talk on the ride home, which was a four-hour drive. I think my friend was stunned or traumatized. We have never spoken of this experience and he has since moved away. I don't think my friend even realized what happened. It took me several years to remember little bits of information myself.

Note: Manzanitas are evergreen shrubs or small trees present from southern British Columbia in Canada, Washington to California and New Mexico. They are characterized by smooth, orange or red bark and stiff, twisting branches. Manzanitas bloom in the winter to early spring and carry berries in spring and summer. The berries and flowers of most species are edible. Native uses of the plant include collecting the berries, drying them, and grinding them up into a coarse meal. Fresh berries and branch tips were also soaked in water and drank, making a refreshing cider. When the bark curls off, it can be used as a tea for nausea and upset stomach. The younger leaves are sometimes plucked and chewed by hikers to deter thirst. Native Americans used nanzanita leaves as toothbrushes.

Snoqualmie Sighting, 1988

Submitted by: Edwin Bonner, age 59, of Hayden, Idaho (retired hypnotherapist, bartender, bouncer).

Sighting: January 1988, Interstate-90 highway, one mile west of Snoqualmie Pass, Washington (Kittitas County).

My fifteen-year-old son Chris and I were headed west on Interstate-90, from Idaho to Seattle. It was around 5:30 p.m. It was snowing pretty good earlier; the snow had eased up somewhat, but the roads were still nasty. We were about one mile west of Snoqualame Pass when I lost control of the car and we spun out.

As I gained control of the situation, we were facing up the bank with the headlights aiming up the bank and into the trees. And there he was, a bigfoot, standing with one arm draped easily around a tree and leaning over the edge of the bank to see what we were doing, with a look like, "What the hell are you doing?"

He had dark eyes, kind of small or maybe recessed a little more than normal. He was covered in muddy brown, matted hair. He looked dirty or unkempt, like he hadn't been down to the river in a while. His neck was kind of short for normal standards and thick or muscular. His shoulders were very broad, but sloped and very rounded.

We figured that he looked to be about seven foot, three inches, and well over 300 pounds—with strong, long arms, I'd say he had about twenty-inch upper arms on him. He continued to watch me arc around the bank and back down onto the highway.

My son and I were pretty somber and quiet for a couple of miles, and then we pulled the car over and talked about what we had just seen and what it was.

We have told people of our experience. Some listen, and some just roll their eyes. But we know. We know what we saw.

Who's Stalking Whom? 1988

Submitted by: Steven Leslie, age 33, of Yakima, Washington (business owner and snowplow operator for Washington State).

Incident: October 14th, 1988, Wahkiacus, Washington (Klickitat County).

This event took place on private land while on a hunting trip in the Klickitat–Wahkiacus area of Washington State. There was a total of seven of us all hunting the same place (two are now deceased). Our family has hunted this area for many years, but on this trip things got a little spooky.

It was the second week of October, opening weekend of deer season. We started off very early Saturday morning, while still dark, driving our vehicle down an old abandoned skid trail in the vicinity of the old Woodruff Mill, to get into a hot spot where we could park my disabled mother on a stand. When it was light enough, my father and I proceeded east on foot, to head off on what we called a "drive" to try to push the deer out of the brush and toward my mother.

The road we were walking down was dense from low fog, and a little soft after a good rain the night before. The area was a pine forest with scrub oak, buck brush, dense stands of mixed conifer, Douglas fir, grand-fir, ponderosa pine, lodge pole pine trees, rolling hills, and gullies. We noticed right away how quiet it was, with no sounds of deer or wildlife, which is very unusual for the area.

We were looking for signs of deer when we encountered strange, eighteen-inch-long tracks right on top of where the tire tracks from our vehicle had rolled through the road. As if something had followed our vehicle. At first we thought it could be a big bear, and immediately we were on alert. But as we examined the tracks more closely, we determined that bears don't walk on their hind feet for this long a period of time, and there were no front paw tracks! These prints had five toes and an arch. They were sunk in the mud at least three inches.

Following the prints, we came to a small draw, at the bottom of which sat a strange structure built out of the surrounding woody debris. If you stood back at a distance, you wouldn't have noticed it, the trees and brush were so thick in there. Brush and vegetation had been piled up into what looked like a dome. In the middle of this dome was an opening where something could crawl in and out.

My father went up close to take a look. I stayed behind,

the smell of cat urine and wet goat was extremely pungent. My father was the only one that was brave enough to go inside. He said it was big enough for three of him to fit in there. We noticed more prints in the area.

We continued on our walk encircling the area. On our way through the thick buck brush, heading west back to the truck, we started hearing noises. Something was now tracking us. We could smell a strong odor, similar to cat urine or a dirty cat box. At this point we were both scared; we could hear it, smell it, but we could not see it.

The visibility in this brush was near five feet in most places, and on the trail we were on, we could see twenty feet to the front or rear of us. We picked up the pace and moved faster toward the truck, and as we sped up, so did it. The brush behind us was loudly cracking so we took cover behind two trees, each of us on the opposite side of each other, aiming our rifles toward the noise. At this time, whatever it was let out a high pitched squeal (almost like a woman giving birth) and took off running northwest of our position.

Hoping it was gone and fearing it was heading for mom, we scrambled that way. When we got to the truck, she was inside with the doors locked. She told us the creature was north of her position and was now making lots of thrashing noise. We went off to investigate, finding more tracks and small, uprooted ponderosa pine trees about four inches in diameter thrown down the gully side, yet no sign of the creature.

We met up with my uncle and cousin, told them the story and showed them the tracks. They were in disbelief. We all went back to our uncle's camp for lunch only to find more tracks in and around his campsite. My cousin is a big guy, his shoe size is a fifteen, and these tracks had more than three to four inches on him. Whatever this thing was, it was big!

My uncle and cousin had spent the night in a canopy on the back of their truck, and had hit the trail early in the morning before the sun came up. We noticed that something had walked around the truck like it was looking in on them while they slept.

We then noticed an old barbwire fence near the campsite, and three posts had been popped out of the ground. We found some grayish/brown hair stuck to the old barbwire and managed to pull up to thirty hairs off of it. My father smelled the hairs and said the odor was so strong that it burned his nasal passage. My uncle then decided he didn't want to stay, he packed his things and left for home.

When we walked back to our own camp, we were not alone. We could hear it moving through the brush as it followed us back. Later we could hear it howling in the distance. It sounded agitated that we were there.

There used to be an old hermit that lived back in this area. He had retired out of the Woodruff Mill. We traded him Yakima apples for his permission to hunt the land. When we asked him if he had ever seen anything strange in the forest, we never got an answer. All he would do is smile.

I don't tell too many people about this for fear of being laughed at or called crazy, but for those of us who were there, we will never forget.

Once You've Seen One..., 1989

Submitted by M.F.D, age 39, of Beaverton, Oregon (industrial sales).

Sighting: Summer 1989, Strawberry Mountains in eastern Oregon (Grant County).

I saw a bigfoot when I was ten years old. Once you've seen one, you've no doubt they are out there. My family was camping with my uncle and his family in eastern Oregon, in the Strawberry Mountains to be exact. My cousin, brother, and I hiked up a stream and we were trying to catch frogs. We ended up at the spring where the stream was being fed.

It was a small area, but had a ring of grass around it for us to sit on.

My brother had to go to the bathroom, and a bush wouldn't suffice for this one, so he headed back to camp and the outhouse. My cousin and I continued to sit there watching the water and waiting for him to get back.

After about twenty minutes, everything got real quiet; there was absolute silence. No birds, no frogs, and literally the hair on the back of my head stood up. We began to hear something rustling in the bushes, maybe thirty feet behind us. I heard a twig snap and looked back over my shoulder to see a dark, leathery looking face, surrounded by dark hair looking at me from over the top of the bushes.

What made me decide to get moving and get out of there was that this creature then stepped sideways, moving out from behind the bushes, thus revealing most of his upper torso and head. I would guess him to have been at least seven feet tall. His head was kind of small with not much of a neck; and his eyes were a dark black. And except for the dark and leathery face, he was covered in dark brown or black hair. He made no other movement. I think he was just as surprised to see us, as I was him.

My cousin heard the rustling, but never saw it. I grabbed his arm, yanked him to his feet and we took off running. Apparently, my reaction was enough to convince him to run like hell too. I never looked back; just ran as fast as I could. We met my brother on the trail and I guess the look on my face said it all. I said, "Run!" and he turned and ran with us. The amazing thing is we were running on slippery rocks and logs, and not once did we fall, trip, or stumble. Adrenalin is pretty amazing.

So, that is my story, and it's absolutely true. I don't repeat it in person because of all the bullshit you get. But like I said at the beginning, once you've seen one, you've no doubt they are out there.

Note: The Strawberry Mountain Wilderness is within the Malheur National Forest in the Blue Mountains of northeast Oregon. Pioneer homesteaders found wild strawberries covered the mountain so they called it Strawberry Butte (or mountain), and that is the name that stuck. Besides the wild strawberries, other abundant vegetation includes, but is not limited to, grouseberries, thin leaf huckleberries, Oregon grapes, swamp gooseberries, thimbleberries, and Sitka alder. Combine this with the twenty-two species of fish in the seven alpine lakes in the wilderness, and what bigfoot wouldn't find this a comfortable and pleasant place to live?

In The Early Morning, 1990

Submitted by: P.J.M., age 51, of Stevenson, Washington (medical professional; drummer).

Sighting: spring 1990, Bridge of the Gods and Stevenson, Washington (Skamania County).

> In the early morning (just before 6:00 a.m.) in the late spring of 1990, I was driving alone between the Bridge of the Gods and Stevenson, Washington (in Skamania County, legendary home of bigfoot, no less!) on Highway 14 to pick up my son who was returning from an overnight school event. The weather was sunny and dry. I was traveling about fifty miles an hour.
>
> Approximately seventy-five feet ahead of me, I saw someone standing on the shoulder of the road in the opposite lane. On that side of the road is a lake, and this "person" appeared to be leaving the lake and attempting to cross the highway.
>
> This "person" was very tall, seven foot or better, and appeared to be dressed completely in dark clothing. I did not have time to observe any other details about this pedestrian, but there were no cars parked on the shoulder or anywhere within sight. I slowed a bit to allow the "person" to cross, which he did effortlessly, and without running, in three strides.
>
> On the opposite side of the road (where he was headed) was a wide, cleared woodland trail, which leads to the Columbia River. I reached the point where the trail inter-

sects the highway within seconds and, as there was no other traffic on the highway, I slowed to peer down the trail to try to see the walker, but he was already gone. This experience felt creepy; I felt my hair stand up, and I did not get out of the car to investigate further.

This "person" was broader at the shoulders than the lower body. For that reason, it impressed me as male. No neck was particularly visible. The "clothing" overall, seemed raggedy and loose, and it was various shades of dark color from head to foot, not at all like separate pieces, such as a jacket and pants, etc. His movements didn't seem natural, and he did not turn to look at me, which seemed odd, as someone crossing in front of a fast-traveling oncoming car would be expected to hold up a hand, or run, or at least look.

I don't know that what I saw was truly a sasquatch, but I don't know any person who can cross a two-lane highway in three steps, and apparently, continue on at that pace!

North to Alaska, 1992

Submitted by: K.R.Tubbs, age 44, of Anchorage, Alaska (CEO of a packaging company).

Sighting: fall 1992, Alcan Highway to Alaska (British Columbia).

First off, I'd like to say that I am the most skeptical person you will ever meet in regards to this stuff. I have spent a lifetime in the bush, and have never seen tracks, or any other sign as may be reported on TV about bigfoot. Further, I think a lot of people are just flat making stuff up, in some sick effort to gain publicity or something.

That said, I have decided to tell you about a trip up the Alcan Highway, completed in the early 1990s. I was driving a new Isuzu Trooper up from Seattle to Anchorage, Alaska for my friend. I was down visiting family and decided I could do that. The Alcan is a long road, and it takes about five days to get from Seattle to Anchorage if you're driving alone. The road will take you through a lot of terrain, from flats through mountain passes and along several bodies of water.

I had been on the Alcan for about a day and a half, and I was about forty-five minutes from the Laird Hot Springs Provincial Parks in BC. The road had a straight stretch that went on for a few miles. On the left side of the road was a sheer drop-off to a steep hill and down to a small stream, about four to five hundred yards.

The mountains are up on the right, making the road about the only flat place for a few miles. This particular stretch had what we call "whoop tee doos." While the road is straight, there are hills big enough to lose the horizon when you are between them, and you can see for miles on the top of the hills. At the lowest part of the road, you can't really see behind you or in front very far.

I was doing about fifty miles per hour and it was about 6:00 p.m. The sun was starting to go down, but I could still see clearly. It had been snowing a little, so most of the ground had a few inches settled, but the road was cleared. When I came up on the whoop tee doos, I was looking far off into the distance ahead of me, and I thought I saw someone on the right side of the road. This would be very strange because this is pretty far out in the middle of nowhere.

Fact was, it was the only thing on the shoulder of the road. I'm sure that was the only reason it caught my eye at first. I would come up on top of the hill and see this form, then lose it when I went down the hill. Then see it again at the top, and lose it again at the bottom.

The closer I got, the stranger this thing looked. First thing I noticed is that it was big; I'm guessing over seven feet tall, on two legs for sure. And it was all the same color, brownish with a lighter tone on the outside. It looked like it was pulling on branches on a tree on the hillside of the road.

I started thinking it must be a large moose, on two legs, reaching up into the trees. Then I'm thinking, "No, it must be a bear…" Each time I got to the top, I tried to talk myself into a logical answer. And each time I went down, I wondered, "What the hell was that"?

When I got to half a mile away, this thing turned around and walked on two legs right across the road and toward the steep cliff. Over the edge it went, and about half a minute later, I was at the place I saw it standing. I had to stop and find out what the hell it was.

I first went to the place where this thing was standing, pulling on the tree, and saw the snow was all disturbed in one place. I could not see a definitive track like a bear paw or moose print, which I am very familiar with. What it looked like to me was that someone had taken a broom, and swept the area in about a fifteen-foot circle.

I went around the car and saw snow swept from one side of the road to the other, but not one definitive track of any kind. This seemed odd to me because this thing was very big. Not mistakable, I mean wide at the shoulders, like a football player in pads, and very tall.

So I followed the snow across the road to the ledge of the other side, and I looked over the edge. It was obvious that the snow was disturbed all the way over the rocks and down to where the drop-off hits the slope down to the bottomland where the trees and stream are.

Now this land is steep; it's quite a ways from the road to the stream, but the land was clear, with not one large bush or tree on the slope. There was evidence of a forest fire in this area, must have been within a few years before my trip. This made the hill down to the water just about clear, until it gets to the area by the stream that was covered on both sides with good-sized trees.

I could see a trail in the snow, like brooms had swept a path all the way to the trees. So I sat on the side of the road and just waited. Years of hunting have trained me that once you have high ground; you can usually get a good glimpse of game if you just hang out a while.

About twenty minutes later, I was cold and hadn't seen anything, when I heard what sounded like a cross between a howl and a scream in the woods below. I played football for a long time, so I think in hundred yard increments pretty well. About 300 yards to my right, I saw the alders flattening, like something huge was moving down the little stream walking on top of the brush.

I couldn't make out any shapes or forms down in the stream bottom, but the alders were being pushed and flattened down in a row. Imagine watching a car slowly drive through a cornfield, but you can't see the car. It was like that. But whatever it was, it had to be very large. Large enough to lay down alders that were a couple inches thick as it cut through the river bottom.

This thing I saw was big. It was covered with the same color of fur or hair, sort of a brownish gray color. No question it walked across the road on two legs. I'm certain whatever it was made the sound—something in over thirty-five years in the woods I have never heard.

I'm not here to tell you I saw a bigfoot, but it is about the best description of what I saw. I couldn't track it, because the path was just not what you would expect. In fact, that's about all it was, a fresh path, with no tracks, if that makes sense. While there were deep indentations down into the snow at what would be about the stride distance you would put your feet if running, the indentations had no particular shape like footmarks. Yet were very much larger than hoof prints.

This was not a bear, this was not a moose. To this day, I have no idea what the hell it was. Some say this sort of yeti, or bigfoot exists, and even after seeing this thing, I still have doubt. But the truth is, it may well have been something I have never seen in the woods, and I have never heard anything like it.

Note: An alder is a type of small tree that grows in spindly bunches, like big bushes, often along the side of streams in this area. They get pretty tall; fifteen feet or so isn't unusual. These were very overgrown, and they must have ranged from a half-inch to a few inches thick on the stalks.

Applegate Lake, 1993

Submitted by: C.L.W., age 25, of Beaverton, Oregon (ticket sales manager).

Incident: spring 1993, Applegate Lake, Oregon (Jackson County).

In the spring of 1993, I was thirteen years old. My family drove out to Applegate Lake, which is about sixteen miles outside of Medford, Oregon. There were six of us, my parents, brother, grandmother, my dog Hay-girl and myself. We were going to hike Collins Mountain, which has a trailhead right off the main road; it goes around the lake and ends at Cooking Greens Campground and Seattle Bar Lake.

We arrived, parked the truck, and began our hike. We were the only vehicle in the lot, and saw no other hikers in the area. I remember how quiet it was, other than the occasional traffic from the main road and the sounds from the creek below. We had only gone about a quarter of a mile in when Hay-girl began crying and headed straight back to the truck. Obviously something was wrong for her to react that way.

I checked out the spot where she got spooked. On the ground about twelve feet off the trail, I found approximately eight huge, human-looking footprints that were well over sixteen inches long. The ground was fairly firm in that area and just slightly damp, yet the prints were about an inch and a half deep in the front and about three inches in the heel. The stride from one print to the next was more than one and a half times that of a human. Whatever it was, it was huge and heavy. I pointed the prints out to my grandmother who remained very calm; she didn't say much, only that she had seen this kind of thing before and she was more then ready to go now.

I visually followed the prints that continued off to the left of the trail down towards a creek/marsh area. We could hear branches and brush loudly snapping in the distance. Obviously a large animal was moving through the brush, and suddenly there was a very loud sound that echoed up the trail. It was a sound like I have never heard before or since, like the sound of a bird crossed with a monkey. It was quick and loud and was gone just as fast as it came. And so were we!

Note: Worthy of note is the reaction of dogs in the presence of bigfoot. Dogs who are normally good watch dogs, will not bark. Those who are normally very brave will whimper or run to a safe place. Even some trained hunting and tracking dogs refuse to follow the tracks of a bigfoot.

Gift of Two, 1993

Submitted by: J.R.A., age 48, of Eugene, Oregon (baker and craft artist).

Encounter: July 1993, Cougar Hot Springs (Terwilliger). Willamette National Forest, Oregon (Lane County).

Many years ago when my sons were teenagers, they wanted to go camping for the Fourth of July. So we decided to head up to Cougar Hot Springs just outside of Springfield. We left late in my small Subaru four-wheel-drive sedan, packed with all of our stuff, five of my son's friends, and the man I was dating—nine total in that little bitty car! We were having a blast, but arrived too late to get an actual camping spot at the Hot Springs, so we decided to drive around the countryside and find an old road that would lead us to an area flat enough to camp.

Finding a nice spot, we set up camp. We told ghost stories and had a great time. My oldest son and one other slept in the car and the rest of us slept outside in a circle under the stars. Of course, leave it to me to be the last to fall asleep as I gazed at the millions of beautifully twinkling night-lights dancing over my head. I heard very few sounds of wildlife; it was very quiet except for the breeze in the trees. I remember thinking that there must be small animals around, but as I listened I didn't hear anything. I was a little nervous, as I truly am not comfortable in the wilderness, and without a tent no less! But I gently fell asleep.

I am a light sleeper, and soon after falling asleep I was startled awake by a rustling noise very close to me. Of course I was frozen in fear, as everyone around me was asleep. I lifted my head, put my glasses on my face, and there in front of me stood not one, but two very clear and definite bigfoot creatures.

They stood not more than fifteen feet from me and it was obvious they saw me. They were completely out in the open, not even a tree to step behind or cover them in any way. The tallest one was at least seven to eight feet tall, and in the moonlight seemed to be the color of dirty white. The smaller one was beige or tan and clearly a young one, not even six feet tall.

I laid there frozen, yet transfixed on both of them. I looked from one to the other as they just stood there. It was dark, very dark, but the two of them seemed to stand out in such brightness. I could see them clearly, because they were facing me. They were thick, not scrawny or underweight in any way. Their faces seemed fully covered in hair; there was no place that was not covered with hair except the palms of their hands and inside of their fingers. Their hair was very long, more than several inches, and very thick.

They had short thick necks; they actually looked like big, hairy, muscular, and very tall athletes. They looked very man like. Their heads were very large, but not ape like, and their eyes were not animal like at all. The taller one had golden eyes, the smaller one had hazel colored on the green side. They both had very, very big feet. I smelled a deep smell of dirt, nothing foul, just like fresh dirt.

The memory of these two outrageously awesome creatures brings tears to my eyes! I could hardly breathe or even blink. They had the ability to communicate and have thought. The little one started to dart off to its left and the larger one groaned at it and gestured with its left arm to stop and come back, and it did. It came back and stood again just staring at me as before. The larger one never took its eyes off of me even when it gestured to the other to come back. The look on its face was both trepidation and curiosity, as it gently swayed side to side.

I was overwhelmed and yet perplexed, but not really afraid of them. I just didn't know what to do; how long would we gaze at each other? I don't know how long we stared like this, but I know my neck began to cramp and I got real cold. So, I did what any red blooded, chicken-hearted woman would do—I took a deep, silent breath and slowly laid my head back down, with my glasses still on so I could peek if I wanted to see if they were still there. I closed my eyes and prayed that I would indeed see the light of day in the morning. For the longest time, I lay there listening for the rustling of the leaves to know if they had left. Never hearing another sound, I fell asleep.

The next morning all the kids thought I was a big storyteller. Sure enough, they didn't believe me. My companion asked me why I didn't wake him as he had a .22 pistol underneath his pillow. I told him how grateful I was that I didn't wake him; would he have shot at them? How dreadfully awful to think! The kids continued to joke and laugh at me when my younger son said, "Hey you guys, look here!"

Someone had wiped a single circle in the dirt covering on the back window of the car, so as to see if anyone was inside. I started to scold my younger son Randy for playing a prank, when Corey, my oldest said, he thought it was his imagination that someone had been peering in the window last night. He thought he saw a hairy face and big eyes. Wow, a sighting that was confirmed—sort of. We did look around for footprints, by the car, and where they stood. The ground was a mess—the leaves and small ground cover (twigs and such) so nothing definitive could be found.

To me, all I need is the eyes of my heart and I will always remember. This memory so overwhelms me; my memory as clear as if it were just last night. As the years blaze by I have found that my precious gift of a sighting has been more of a gift than I could ever know it would be on that night. I rarely tell my true sighting to people, but once I told it to my sister, whose American Indian friend said that I am a truly blessed person as bigfoot rarely allow themselves to be seen, let alone two of them at the same time to the same person.

The Tanner Creek Stalker, 1993

Submitted by: Lex Loeb, age 49, of Cascade Locks, Oregon (office manager).

Incident: summer 1993, Tanner Creek valley area of Oregon (Multnomah County).

I had an encounter in the summer of 1993; I was running a trail, or rather a service road, behind the closed gate at the top of the Tanner Creek area. I was enjoying my run and my solitude when suddenly something began pacing and following me swiftly and deep in the bush down below the trail. And it was talking. I was a good trail runner at the time, and was running an open trail; I can't imagine how any person could be running as fast as I was through the thick brush and below on those steep mountain sides.

As I ran faster, it ran faster, and continued to pace and stalk me. I called out to whatever it was and it mocked me. I called out, "Who's there?" and got back something that sounded about the same. This terrain is not an echo chamber. I then said, "Hello, hello, hello" and got back "Hi yo, hi yo, hi yo!" I ran the hell out of there much faster than I normally could have to my car and got out of there.

It may be the reason the forest service shut off access up there. It's probably a half-day or whole day hike now, since they closed the road, but if you want your own experience I suggest you try going up there. You just take Interstate-84 to the Bonneville Dam exit and instead of going under the overpass to the dam you head toward Tanner Creek trail and

Wahclella Falls. Turn to the left and go up the dirt road going up the mountain side, as far as they now let you go up the road (or you may have to hike in now). The service/fire road is still there, but I think they closed it off. Up at the top there is a big forested wilderness between that and Mt. Hood. You probably have to hike up above the 3,000- to 4,000-foot elevation to descend into the Tanner Creek valley. It winds back toward Mt. Hood and is a veritable wilderness, save some high tension powerlines.

From the top you see nothing but douglas fir and western red cedar cover along the creek and in the valley deep below. It's not Mura Point that you are interested in, but the deep, dark forests in the upper plateau valley that wind around with the river toward Mt. Hood, which is an active volcano. You'll find the Basalt Cliffs where there are many caves hidden up there.

Skamania County, with its annual bigfoot festival in Stevenson, is directly across the Columbia River and dam from this part of Oregon. There are some amazing geographic features there across the river, including the lava beds, which look like a giant Japanese garden with stunted bonsai plants growing in the crevices. One can easily get lost in there and compasses don't work. The ground below your feet seems hollow. There are many jagged lava caves there. It has to be seen to be believed.

You may not know it, but the Indians had placed their most sacred Bradford Island graveyard right where the dam was built. The ancient Indian graveyard was destroyed to make way for the hydroelectric dam. I actually believe the Washington side of the river is more fascinating at that point in the river, with beacon rock, natural hot springs and evidence of violent rockslides, so powerful that they dammed the river creating a land bridge. You can Google the site and maybe find photos. Hardly anyone goes up there since the forest service closed off the road access.

Fast-Moving Sasquatch, 1994

Submitted by: Dusty Dornay, age 31, of Seattle, Washington (mechanic by day; bouncer by night).

Sighting: September 1994, Mt. Spokane, Washington (Spokane County).

> A friend and I were bow hunting on Mt. Spokane up by the ski lifts and "Gussy's little restaurant." The day was warm and clear. We both saw a creature outlined on the ridge above us, probably at 1,000 yards or better. It was an upright figure at least a foot or two taller than a "tall guy," with a fairly wide body. We could not make out any features as the sun was behind it on the ridge, but it caught our eye right away because of how fast it was moving.
>
> At first we thought we spooked an elk, but right away noticed it was upright and moving on two legs the whole length of the ridge to the tree line (probably 5,000 yards). And when I say fast, I mean probably thirty or forty miles per hour; it was as fast as I have ever seen an elk or deer cut loose. There is no doubt in either one of our minds as to what we saw. It actually kinda ruined our hunt and needless to say, we haven't hunted Spokane County since. This is a no "bull" story and I swear by it.

Note: The following information on "sasquatch speed" is from a paper written by Dr. W. Henner Fahrenbach, entitled "Sasquatch: Size, Scaling, and Statistics."

> The maximal speed that a sasquatch is capable of attaining has not been reliably tracked, although many casual reports refer to observers driving in a vehicle parallel to a running sasquatch or a sasquatch running with galloping horses. Before rejecting unbelievable sounding speeds or step intervals, it is well worth keeping human records in mind. For example, the world record walking speed over 20 km is about

> 11 m.p.h; the top running burst speed is about 27 mph, the longest single jump near 30' and the longest triple jump—in effect three running steps—about 60', all this with a physique of decidedly smaller scale than that of a sasquatch. Extrapolation from step lengths and observed running cadence (about 140 steps/min.) suggests a top speed of the sasquatch of 35-40 miles per hour, the speed of a galloping horse.

The full, unabridged version of Dr. Fahrenbach's paper is available at: www.rfthomas.clara.net/papers/size1.html

Watched on the River, 1995

Submitted by: Melinda Mae Reynolds, age 32, of Kelso, Washington (dedicated homemaker and mother of a large beautiful family).

Sighting: August 1995, Abernathy Creek, Washington (Cowlitz County).

> I spend a lot of time in the woods hunting in the great northwest. At this time in 1995, I was not hunting, as I was four months pregnant. Instead, my boyfriend and I went camping along the Abernathy Creek in Cowlitz County, Washington. This area was downstream from a fish hatchery and almost to the mouth where the creek dumps into the Columbia River.
>
> We had our tent on a flat spot about ten feet from the river. We had a small campfire and were sitting around talking and enjoying the night, when we began to smell something bad. The closest thing I can compare it to is an elk during rut (mating season); they have a very strong, musty odor.
>
> Suddenly, we heard something in the brush across the river on the other side of the bank. It was only about eight feet across so we shined our flashlight over and saw nothing. I assumed it was probably a deer going down for a drink.

About twenty minutes later, we could hear branches breaking and loud crunching. The noise became more intense by the minute, and from experience I knew that was no deer! I got a very unsettling feeling and again shined the light across to the other side of the river.

There was a lot of thick brush and trees, but animal eyes "shine" when a light hits them. Suddenly, something did catch my eye and I froze in disbelief. I saw a furry face looking back at me. I guess it looked like all the bigfoot drawings I have seen on TV, but this one had dirty white hair, not dark.

His arm was stretched a little above and to the side; his hand was up on some tree limbs as he was pushing them aside— almost as if he was holding a curtain back to look out a window. All I could see was from the head to just below the shoulders, but it was very tall. Where he was standing on the other side was above the riverbank area on about six foot of higher ground than we were.

Given that, my guess would be that he was approximately seven feet tall, although I got the impression he was bending to look through the opening in the brush. The hair was longer around the head area than the shoulders and arms.

Circling the face was all hair, yet the hair was sparse around the eyes and nose area. The hair seemed to stick up rather than laid or brushed back—about four to six inches long. Kinda' like those funny looking lizards that have the frill around their head.

His mouth was closed, almost drawn down, so I saw no teeth. The nose almost seemed like an ape with the wide nostrils. I almost swear the eyes shined back as red, but that might have been the fear that made me see that. I know deer and other animals' eyes shine yellow and green. He was not doing anything but looking at us.

We threw sand as fast as we could to put out the fire, left the tent and got out of there quick. Sheer fright would be a good description of how I felt. My boyfriend did not see it, but he heard all the loud noises in the brush and knew that I was good at identifying animals, and this was one animal I could not identify.

I have been tight lipped about this thing, knowing I would get teased, and have never told anyone. I have hunted everything from bears, cougars, raccoons, deer, and elk, and have never in all my time seen anything stand upright and look like this did. I do not know if it meant us harm, but if it did, it had ample opportunity, and did not. It just looked at us.

I have heard two other people talk about almost the same creature in close proximity to this location. I can still see the face, I can still feel the fear, and will never ever forget for the rest of my life. That is what I saw, and I am a God-fearing woman.

Bigfoot Followed Me Home, 1996

Submitted by: M.J.H., age 49, of Portland, Oregon (professional test driver for Chrysler Corporation; currently drive tests 2012 engines in the custom designed prototype vehicle of the future).

Encounter: summer 1996, Sandy River at Oxbow State Park, Troutdale, Oregon (Multnomah County).

I was camping with my son, on the Sandy River at Oxbow State Park. I have been going there every summer for twenty-plus years. In the morning after breakfast I took my son for a walk through the woods. He was four at the time.

We had walked only a quarter of a mile when we came upon a ravine overgrown with vines and trees. I heard a loud crack and turned to see the ravine being parted by something very large. This was about 100 feet from where we were and the ravine was almost vertical, very steep and impossible to climb unless you wanted to cut away an almost jungle like growth that was about sixty to eighty feet high.

I thought it was a bear or a wild boar, or maybe a giant

deer. (I learned later from a park ranger that there are neither bears nor really big game in the area.) I was frozen in fear for almost five seconds and couldn't believe what I was seeing. The foliage was loudly and violently being pushed apart almost seven feet across. Then a smell like a dead animal mixed with cat spray and a musk odor just overwhelmed me. It was so strong that if I hadn't been so scared I would have thrown up!

I picked up my son who was stunned and staring wide-eyed behind me, and took off as fast as I could. I went back to our campsite and packed everything. Before leaving for home, I took my son down to the river and let him play most of the day just to stay away from the forest behind my campsite. There were a lot of people playing in the water. I was more scared than I had ever been in my life, and not having anyone to tell, and not really wanting to for fear of being teased.

The next day I stopped by a small cabin in Troutdale, which I rented from an old man. I never unloaded from the day before, so I stopped to put the camping gear away. When I got back in my truck, I noticed a horrible smell on the outside and inside of the truck, but thought maybe I had hit a skunk or ran over some old road kill. I ignored it and headed home.

Two weeks later (during the evening), I went back to the cabin to get a toy train set I had hidden for my son's birthday (this was around ten at night, after I got off work). I went inside, grabbed the train set, and that horrible smell hit me again. I started to shake and was so frozen with fear I could hardly breathe. I opened the door, my truck only about four feet away, and as I ran to it and threw the train set in the front seat, this incredibly loud, shrill-like scream hit me full blast from behind the cabin about three feet away. Then a gurgling sound and then another scream!

I jumped in my truck and tore out of there. About six car lengths away I looked back to see if I could see anything, and sure enough there was a bigfoot standing right where my truck had been parked! It was about eight feet tall with matted light-brown, almost tan-colored, long hair, maybe seven inches long. He had no neck to speak of and I remember, as he was standing next to the cabin under a light, that his arms hung almost to his knees. He was just standing there looking right at me. I have never been so terrified. I floored my truck and didn't stop until I reached downtown Gresham, about ten minutes away.

Don't ask me if he followed me from the park or just lived nearby, I'll never know. I do know that the cabin sits right on the same river and backs up to the same land as Oxbow Park, but the park is fifteen or so miles from the cabin.

About three months later I had a friend help me empty out the cabin. While moving my stuff out, the old man who owned the cabin showed up. I "joked" with the old guy that I had seen bigfoot. He said, "Oh hell, they have been here since my great granddad used to run choke chain at the old mill up river!" He was a serious type of old timer and didn't talk much, so I didn't go into any detail with him.

I hope I never see it again, as it was something evil. I don't know how else to describe it, other than to say I felt a presence both times that was one of complete terror and doom of which I had never before nor since experienced. I am a very caring person and love all animals and creatures, and I have never had this feeling about anything in my life. I don't think it was a "normal" one. Hell, before this ever happened I would have gone on a trip to look for them, but for some reason this one was different. And I know it was the same one that came down that ravine; I don't know how to explain it, but I know it was the same one.

To this day I won't go camping and sleep in a tent; I have to be in a motor home. And I don't go exploring or hiking. I swear this is a true story, and would even take a lie-detector test to prove it.

A Close Call, 1996

Submitted by: Frank O'Connor, age 37, of Salem, Oregon (professional fence installer).

Sighting: late summer 1996, Interstate-5 freeway on the Siskiyou Summit, Northern California.

I travel to and from California to Oregon several times a year. In the summer of 1996, around midnight, I was driving my 1968 Chevy El Camino, heading north on Interstate-5 at the Siskiyou Summit. The roads were clear with stars in the sky. I was rounding a corner and came across the top of a hill, when I saw a figure jump over the center divider. I thought it was a person, so I immediately hit the brakes. The figure stayed in motion, not running, but walking.

He stopped almost on the centerline dividing the slow and passing lanes. I swerved and stopped pretty much in the middle of both lanes. With my car in the middle of the road, not more than twenty feet away from him, I could see plain as day that it was a bigfoot. This was not a sighting from a distance; I came within feet of hitting him, and he looked me right in the eyes. We sat there just staring at each other.

He was very large, over seven feet tall. Tall like a basketball player, but built like Paul Bunyan. All around built tough. He was kind of hunchback. His hair was a medium dark brown and covered his body, about three inches in length and very unkempt, just a hairy mess really. His arms were incredibly long. His eyes were dark and matched his coat, they just blended. I don't recall any skin like I have seen in drawings; he had hair all over his body and face. He looked to me like a large human with hair all over. If you have seen the movie *Harry and the Hendersons,* he really looked almost like that. The build, the hair and shape; it was like the person who made the character had seen a bigfoot himself.

We looked at each other for what seemed an eternity. I was really trying to believe what I was seeing, and I looked for as long as I could. The creature seemed just as curious about me as I was him. He looked like he was scared, unfamiliar with the territory he was in, and very hesitant to leave. I do not remember being scared; I was well aware of his enormous size and strength, but I never felt threatened. He finally started walking again, a few long strides and he was off to the side of the roadway and gone.

Frank O'Connor's drawing of the creature he saw.

I had the chance to try and run him over; for the sake of science I thought of it—hitting him and capturing him. Had I not braked in the first place, had I not been paying attention, there was a great chance I would have hit him accidentally. I realized that I had come across a very rare opportunity to actually see this elusive creature firsthand. And he is better left alive and alone, to dwell in his own environment. I always believed in bigfoot, and now I am a *true believer*. I have seen him, I can tell you it was not a hoax, and he does look like all the pictures you see around.

Mysterious Footprints, 1997

Submitted by: Ellen Rubenson, age 43, of Ashland, Oregon (medical case manager and writer).

Incident: In the early spring 1997, Ashland, Oregon (Jackson County).

Tucked in a valley at the edge of the Siskiyou Mountains lies the Shakespearean town of Ashland, Oregon. Tourists flock to the bustling downtown galleries, attend Shakespearean plays, and stroll through beautiful Lithia Park along picturesque Lithia Creek. Above the park, the locals mountain bike, walk their dogs, and hike the numerous trails.

Named from the Lewis Carroll novel, *Through the Looking Glass,* the trails hold names like, Alice in Wonderland, Caterpillar, Rabbit Hole, and White Rabbit. It was on the lower half of White Rabbit, just above the Oredson–Todd Woods and the North Mountain Park that I saw a footprint in the dirt by the side of the trail.

I love the mystery of bigfoot and I love the idea that bigfoot exists. I have even visited the bigfoot trap off the Collings Mountain Trail in the Applegate Valley to place my hands on the heavy metal grated door and peer inside the dark, musty, wooden cage designed to ensnare him. I can only imagine the locals quietly manning it, patiently waiting for a hairy bigfoot to reach for the bait planted inside, thus springing the trap.

I can visualize their excitement as they point their tranquilizer guns at him and then watch, as he slowly drops to the ground, asleep. It is reassuring to know that the locals did not intend to kill or hold the bigfoot captive; their intent was to attach a transmitter to be able to track him.

The footprint rested on the edge of the trail. It was at least eighteen inches long, six inches wide, and had distinct imprints of toes. Down the hillside, I saw a second print. The stride between the two footprints easily stretched four to five feet.

The following week, I attended a bigfoot exhibit at the Science Museum in Ashland. As I paid my entrance fee, I noticed a questionnaire asking, "Have You Ever Had A Bigfoot Sighting?" Curious, I completed it and then walked into the exhibit. I read about bigfoot's history, viewed videos of real bigfoot sightings, and eventually found myself in front of a description of a bigfoot footprint. To my amazement, and the amazement of a complete stranger next to me, I found myself blurting aloud, "I can't believe it. I actually described a bigfoot print!" I felt stunned, and it affirmed my belief that bigfoot does indeed exist.

Upper Cispus River, 1997

Submitted by: R.T.G., age 44, of Eatonville, Washington (former firefighter/EMT– medical sales arena).

Incident: September of 1997, Upper Cispus River drainage area, East of Packwood, Washington (Eastern Lewis County).

My friend and I were out scouting for deer in eastern Lewis County, in the mountains that border the Goat Rocks Wilderness. We made up our camp, and it was just after dusk when we settled in to enjoy a nice fire. It was so far an uneventful early evening, a usual dusk in the hills.

It is usually very quiet up there, and aside from the occasional booming of a great gray owl, there is rarely much noise in the early fall. We had our rifles with us, stored in the truck. Directly below our campsite was an old, overgrown clear-cut, surrounded by large second growth trees. The clear-cut was extremely steep and rugged.

All of a sudden, chaos erupted when something started to rip things apart down the slope from where we sat. It was an animal of substantial size based on the damage it was inflicting. The sounds were similar to the sound an elk makes when raking a sapling, yet much huffing, grunting, crashing ,and some heavy thudding, as if heavy things were being thrown about.

Suddenly, it let out this scream that made our hair stand on end. It was a guttural scream that wavered. It was close enough, roughly 200 yards away, that you could almost feel it. This screaming continued for several minutes. Then it went dead quiet. Very quiet.

By that time, we scrambled to the truck to secure our rifles. You have never seen two grown men so scared. We have been roaming the backwoods for many decades, but have never heard anything like this. We have heard bear, elk, deer, cougar, bobcat, and elk bugle in my yard. They rut in my trees, but this was not even remotely close to any of those animals or those sort of sounds. More like a primate you would hear at the zoo. Almost guttural huffing followed by a rising cry, like a woman's scream at the end.

We stayed in the truck the entire night and left at first light. We could not see any disturbance below the camp when the sun rose; but then again it was very rugged. To this day, my friend and I will not camp in that area. We have never been back.

My father reminded me of a fishing trip we took together many years ago (in the early 1970s) when we hiked into Chambers Lake, about three miles above this same campsite. The road was snowed in and we had hiked six miles to the lake. There were neither footprints from other people nor any ski mobile tracks; we were alone at the lake.

On the far side of the lake is a shale slope that rounds a small hill. As we fished, something was moving on the slope as we could hear the shale sliding. Suddenly it stopped, and from the tree line came this same unearthly cry that was nearly identical to the cry heard with my friend, though not as forceful. We could hear it moving on the back of the slope, just out of view. I am certain we could have seen it in the tree line with glasses.

My father became very scared, grabbing my arm and telling me to leave the fish. He grabbed the rods and almost dragged me all the way to the road. He had his pistol out and kept looking back; he nearly walked backwards all the way to the car. When we made the road, I noticed that the rods still had their line out and we were just dragging them behind us. I remember the worry and fear he had. We never went back to fish that lake.

Having since heard the recordings of sasquatch, my friend and I are both certain that what we heard is the same sound mentioned in other sightings. And after doing some research on the area (between Mt. Rainier and Mt. Adams) as well as the Dark Divide, there appears to be a historical record of sasquatch in the area. I am certain they existed prior to the white man's entry into the area.

The Native lore is rife with references. Some valleys in British Columbia were off limits due to reported sightings. I do not think they are some kind of monster, but more likely a boreal primate. Much like a gorilla or proboscis monkey that can tolerate cold weather.

I remind people that the great panda went undiscovered for a long period of time, as did the giant squid and gorilla. It wouldn't be a stretch to have a secluded primate in small numbers roaming the backwoods. Stranger things have happened.

Band of Bigfoot, 1997

Submitted by: J.J.S., age 46, of Buckley, Washington (electrical engineer).

Incident: November 1997, Trapper Creek wilderness, Washington (Skamania County).

I grew up in Carson, Washington, and lived in the surreal beauty of the Columbia River Gorge and the heart of the Gifford Pinchot National Forrest for twenty-seven years. As a younger man, my friends and I would hunt and fish every chance we got. We were accustomed to all the creatures of the forest and knew most of the areas like the backs of our hands. A lot of famous bigfoot sightings have been in or around this area of Washington State.

I actually lived next door to Datus Perry, who claimed to have seen the creature on many occasions. And I was also friends with Louie Alway and his daughter, who stated their sighting on a national TV documentary on bigfoot. Old Ray Bleigh used to tell my dad and I that a circus train derailed near the Carson/Stevenson area in the early 1900s and some apes escaped the wreckage and mated with bears, and that's what everyone keeps seeing. Their stories, as they stated them to me, seemed...well, let's just say a little unbelievable at the time. You see, up until my encounter, I myself had never seen or heard anything remotely like that in the big woods. Now for my story.

It was November, and high-powered rifle season. I had planned on elk hunting with my brother-in-law, as I almost always did. This particular day, he had other responsibilities and could not go, so I decided to go by myself with the intent of not going to the far off places we usually hunt. Instead, I went hiking on a horse trail in the Trapper Creek Wilderness, a forbidding place where man usually only goes on horseback. There are miles and miles of straight up-and-down canyons and old-growth fir trees blocking any sign of where you are without a compass. A place uniquely untouched and un-trekked by man.

In my excitement of finding some very fresh elk tracks, I wandered off the horse trail, which was an insane move, because that was my sure path back to my truck. My only redeeming forethought was that I had found an old creek bed and decided that, if lost, I could follow it back out of the canyon to the horse trail. This creek bed was perfect for hunting; it had huge boulders covered in three inches of thick moss, so I could step from boulder to boulder without making any noise at all. I was so quiet, and amazed at my stealth, before I knew it I had gone in a few miles.

All of a sudden, the hair stood straight up on my neck and arms, and I thought, "Uh oh, I've felt this feeling before." It's usually when a cougar is watching you from a tree or a den that you have wandered too close to. I stood as still as stone and listened intently as I clicked the safety off on my rifle, expecting a full frontal attack.

The area on both sides of the creek bed was heavily brushed, with a thick moss base on the ground. Ten feet of sight through the brush was seeing a long ways. Suddenly, I got several strong whiffs of the foulest smell I have ever smelled, like that of a dead and rotting animal carcass. I could hear shuffling all around me in the bushes and the trees; it was as if the woods came alive around me.

Standing still and quiet, it became obvious to me that it was not elk, cougar, or bear that surrounded me. These were creatures unknown to me. It didn't take long to realize that I had snuck right up and into the middle of a whole damn family of bigfoot.

They could see me, I knew, but I could not see them. They continued to shuffle through the bushes, but were very careful to stay hidden from my sight, like ghosts in the woods. I was scared, very scared. I decided to slowly turn and start back up the creek bed the way I had come. As I turned, I heard a sound that I have never before heard in the woods or any place since. It was like a high-pitched whipping sound, as if you took a long rope or cable and twirled it around as fast as you could.

I stopped dead when I heard the first one, but then I heard another off in the distance, and then another one real close to me. I started almost running up those boulders. As I moved through the creek, I realized they were following me. I would look side to side and not see any shapes, but I could hear them moving quickly through the brush, yet smoothly, as if through a dense cornfield. It would take an enormous amount of strength to walk through that brush at that speed. No doubt as to what it was now, as I knew they were running on two legs to keep stride for stride with me on a parallel direction on both sides.

There were at least three of them. One on my right and two or more on my left. As I continued to move, the whirling sounds would get louder or quieter depending on where I was in the creek bed. It was as if they were communicating my position or something. I couldn't just hop from rock to rock in a straight line; I had to go left or right to get around stumps, downed trees and foliage. When I veered left, the whistle-swooshing-whirling sound got louder. When I strayed back toward the middle of the creek bed or toward the right, the sound was more faint.

I gathered that the left side was definitely not where

they wanted me. They stayed with me for a good quarter of a mile or so, the longest quarter-mile run of my life, until they made sure I had entirely left their domain. I never stopped until I reached my truck.

What the whirling sounds were, or how they were made, I don't know. In all the years I've spent in the big woods, I've never heard anything like it. That is the only time I have ever been literally scared in the woods. And I had a high-powered rifle with me. I have never returned to that location, and I don't want to.

Note: Possibly to scare people out of their territory, bigfoot has often been said to run or chase alongside, though out of sight, only stopping when the terrain would expose them to view. On other occasions they have been known to throw rocks, shake trees, or repeatedly break large sticks or branches (thereby making lots of noise). These aggressive displays are also found in known species of great apes.

The Cabin Creek Campground Visitor, 1998

Submitted by: Scott Coats, age 40, of Kenmore, Washington (territory manager for a large chemical company), and Amanda E. Oliveira, age 36, same location (city planning inspector).

Sighting: summer 1998, Cabin Creek Campground, Wenatchee National Forest, Washington (Kittitas County).

My wife Amanda and I decided to go to the Cabin Creek Campground, mid-summer, to enjoy an outing with some friends. They have a trout lake that you can fish on, and they charge by the pound. All the good cabins were pretty much taken up, so, being outdoors people, we pitched the tent.

We stayed up with our group of eight people, had a fire ,and a few adult beverages. The next day we got up, had breakfast, and then decided what we wanted to do. My wife likes exploring the logging roads and looking for wildlife. We have seen deer, elk, and bears—even a lynx!

We got back to the tent at around three-thirty Saturday afternoon, and busted out our lunch. After we ate, I decided to go fly-fishing on the river about six miles up the road from the campground. The rest of our group was either fishing the pond onsite on just kicking back. So I packed up my gear in the pathfinder, and asked Amanda if she wanted to go for a few hours. She immediately grabbed the book she was reading and jumped in the car.

We got up there and set up shop. She laid down a towel and sat and read while sipping some wine, and I went down the river to fly-fish. Amanda, keep in mind, was 200 feet from the car. I came back up the river after about an hour with no luck of fish. I sat down with my wife, took a break, and had a beer and some water.

After about fifteen minutes, I decided to give it one more try. On the way up, I saw a good fishing hole about 1,500 feet up and headed in that direction. It was now around 8:00 p.m. and the sun was setting below the hills. Still light out, I knew I had about an hour before we should pack it up and head back to camp.

I walked the river and was casting back and forth, when I started hearing these loud clicking noises. I looked around, and did not see anything. I walked about 100 feet further up the river, still fishing, and began to hear more noises. Still the clicking sound, but louder and very close! At that point, I got the feeling that someone or something was watching me. I have been an outdoorsman all my life and have never heard a noise like this. I started walking back to the car, reeling as I walked, and heard more noises!

I thought, "Okay, I have no clue what the hell is in that brush, but I am not about to find out." Plus, I was unarmed. I got about 300 feet from my wife when all hell broke loose. There was a loud crunch; I didn't look back, running full throttle to the car. When I got to the car, I looked back to see this thing going up the side of the hill like a raged ape. By the time I saw it, it was about 300 feet up the hill, going at least twenty-five miles per hour.

I was a short distance away, but I could clearly see that it was kind of human like. It was maybe eight feet tall. The hair was very dark brown. It had no neck. It was no bear, and it was huge. It looked like a human on some serious steroids, very muscular. If I was to climb that same hill it did, it would have taken me five minutes at the least, and I am in very good shape. This thing did it in about twenty-five seconds.

I was so overwhelmed that I started the car, and damn near left my wife on the riverbank! She came up yelling, "What the hell is going on?" She said my face was as white as a ghost. Then she saw it. The thing had got to the top of the hill and we saw it crouch down on one knee, about five hundred feet up, watching us. I was blown away!

The uneasy feeling I felt when I heard the noises, I have never experienced before, ever. I was scared, but I think it was just as scared as I was. We actually packed up the tent that night, and got a cabin for the rest of the stay.

Scenic Highway Strange Sight, 1998

Submitted by: R.S.W., age 35, of Portland, Oregon (firefighter and high rise window washer; master wood craftsman).

Sighting: summer 1998, Columbia River Scenic Highway, Oregon (Multnomah County).

It was late afternoon, early evening, in the summer of 1998. I was headed up the Columbia Scenic Highway to my girlfriend's house in the little town of Corbett. The road is pretty curvy, and as I rounded one corner, I saw a huge, hairy creature walking on two legs moving quickly into the woods to my right.

It was about seven feet tall with dark hair. At first I thought I might be seeing things, but when I got up to where it was and stopped, I smelled the foulest stench I have ever smelled. I have not smelled anything like it before or since. My friends who were following in a car behind me stopped and they could smell it too.

We went back later to look for prints or hair samples on the bushes, but found none. After telling this story to others I found out that a smell is sometimes associated with bigfoot and that I probably had a genuine sighting. There's no question for me as to what it was I saw.

I know that our forests are so big that nobody realizes what's out there. The tribes in the Congo of Africa claim they see long-necked creatures that resemble small brontosaurs that eat plants in the swamps.

We can only go as far into the jungles as the food we can carry; thirty days food is all you can get in there, so that's as far as we can go. The Congo is 50,000 square miles; the Pacific Northwest is even more.

Rogue River "Rogue," 1998

Submitted by: J.L.M., age 49, of Renton, Washington (private duty nurse).

Sighting: summer 1998, Upper Rogue River, Southern Oregon, (Josephine County).

I grew up in the mountains and in the country, and I have seen and heard bigfoot many times. This is one of those times:

It was the summer of 1998. My husband, my two young daughters, my son, and I were camping in southern Oregon on the upper Rogue River, just beyond Indian Mary Park. We needed some groceries and snacks one day, so we hiked the two miles on the newly paved road to the Galice general store. On the way, we met a man named Randy and his wife who were camping close by, and walked with them and their two boys.

When we were about halfway back from the store,

Randy's boys ran ahead, racing each other back to camp. It wasn't long before one of the boys came running back white as a sheet. He kept pointing down the road toward camp saying, "Big monster, big monster."

We took off running in the direction he was pointing and found the other boy standing frozen and pointing down the side of the road into a blackberry patch. The boys said a large, hairy monster had ran across the road in front of them, stepped over the guardrail and headed down the bank.

We noticed several footprints. One distinct print was in the fresh asphalt and gravel that was packed down on our side of the newly installed, three-foot high guardrail. The print was over twenty-four inches long and at least fifteen inches wide. The toes were bent from the little toe to the third toe. We could also see that it had stepped over the guardrail, as two other prints were deeply indented in gravel about three feet apart and heading down the bank into the blackberry bushes.

The bushes were about ten feet tall in places, but we found a distinct path that went about forty to fifty feet into the bushes. We followed it and that's where we found its bedding-down place hidden in the midst of the blackberries, along with a patch of hair that was caught on a blackberry stalk. The hair was about three to five inches long, was a reddish/brownish-black color with a putrid smell and it had some skin attached to it. The smell was very bad, like nothing I have ever smelled before, and I grew up in the country! We placed the hair in a Ziplock bag to keep.

We walked back to camp a bit beside ourselves, wondering what we should do; needless to say we didn't sleep much that night. We calmed the kids best we could and built a huge camp fire that we kept burning all night long. Come morning we talked to others who were camped further down the way, and about five families moved closer to our group. We took turns staying up all night and the next night as well. Randy did have a big dog and this helped as he always let us know when someone or something was near our camp. Or so we thought.

Late on the third night, as we sat around the campfire, I looked about 100 feet out across the river, and in the moonlight I saw him: bigfoot. He was standing straight up and had to be over ten feet tall. Walking like a man, he moved very quickly straight up the steep side of the mountain. He was dark in color, his color kind of blending in with the rocks.

Two days later and just at dawn, Randy and his family, who were now camped not more than ten feet away on the other side of some bushes and small trees, came over to tell of an unnerving experience. Randy was quite upset and asked who was playing tricks last night. He said it wasn't funny and thought it was a dirty trick. Confused, we asked what he was talking about. A funny look came over his face as he said, "You mean it wasn't you?" He then began to describe his visitor during the night.

They had retired into their small tent for the night; their dog was outside tied to a long lease. At some point during the night, without making a sound, the dog managed to wiggle through the zipped opening of the tent and began to shake and whimper, refusing to leave. They unhooked him and he immediately crawled into one of their sleeping bags, going all the way to the foot. As they settled down to sleep once again, a large hand appeared on the outside of the tent and began to push down on the top of it. Randy watched as the hand pushed so far down as to almost touch them.

Randy, who was a big guy, a drummer and quick with his hands, reached up and grabbed the thing around the wrist. He said it pulled away with such strength it was as if nothing had a hold of it. He then demonstrated by having my husband lean over and reenacted the incident. My husband was six feet tall and 360 pounds, by no means a weakling, but when Randy grabbed his wrist, my husband found he could not pull free.

Several footprints were found outside his tent as well. Sand surrounded his camp area but we could make out where something had been walking around the site and around the tent. The impressions were deep, so whatever it was, was big and heavy.

We never heard a thing. We did however, notice how eerily quiet it was that night before; not even a cricket was heard. Growing up in the mountains of that area, I remember the old timers reminding us kids that when the woods get real quiet, you climb a tree and don't breathe. And you stay there till you hear the crickets again, and then it's safe to come down. I guess they were right. We all gathered together in a group after that, about five families all camping close together.

Before we packed up and left our camp, we took the hair samples to the Grants Pass Forestry Department. There were two men in the office; we gave them the baggie and told them our stories. They were polite and thanked us; said they'd know what to do with the sample and that it might be best for all concerned if we forgot about it altogether.

You can never forget. Like I said, I grew up in the country. I've seen panthers, cougars, wolves, and bears up close. I've seen it all. Now there is something else I've seen—a bigfoot. They're real.

Bigfoot in the Road, 1998

Submitted by: A & G.H., of Anderson, California (owners of a specialized and oversized trucks trucking business that serves forty-eight states and Canada).

Sighting: September 1998, Highway 38 to the Coast of Oregon (Douglas County).

In 1998, my wife, three boys, and I were traveling from Redding, California to Seal Rock, Oregon on the coast to visit my father and stepmom. While traveling on the highway it was already dark and there were many twists and turns. Rounding a corner, I noticed an RV parked precariously on the right side of the road, as though it had stopped in a hurry. People were milling around from other cars, which had stopped as well, so the area was lit up with headlights pretty good.

I turned my attention back to the road to see a creature with the stature of a man sitting on the road directly in front of my car! I turned sharply to avoid hitting it and began to slow down. I was about to turn around and go back, when my youngest son, Dustin, cried from the back seat, "Please don't stop, keep going! I'm afraid."

A chilling picture began to develop in my mind. It looked like a human but was covered from head to toe with long brown and gray hair. It had no clothing on and was sitting on the road, and had probably been struck by the RV we had just passed. My god, someone just hit a bigfoot and we almost hit it too! After much pandemonium, we concluded we had just witnessed a bigfoot, and did not stop until we reached the coast and found a grocery store parking lot to pull into.

It was obviously different from anything we've ever seen before. Even after seeing the creature many miles back, my youngest son didn't want to get out of the car. I stand five feet, eight inches tall, so I went and sat in front of my car to try and gauge the stature of the creature. Its head and shoulders were above the hood of my car, sitting square on the ground. My head barely reached the hood.

Soon after, my wife went to the Wax Museum in Newport, Oregon where there was a wax bigfoot. She said it had an uncanny resemblance to the bigfoot we saw.

Clack-Clack-Scratch, 1999

Submitted by: L.C.C., age 61, of Washington State (registered nurse.)

Incident: autumn of 1999, on an Indian reservation in northwestern Washington State.

I was at a weekend spirituality retreat at a friend's home in Washington State (exact location not to be divulged), who happens to be a Native spiritual leader. It was a two-night retreat and I asked if I could put my tent in the trees. A place was created for me through a little twisted path that opened onto a small clearing. I pitched my tent in the very center. No trees or branches were touching my tent and you could walk all the way around it easily.

While I was within shouting distance of the Leader's home, I wasn't within sight. And being as I'm female and would be sleeping alone, I was sent to bed with a large bell that I was told to ring if I had any trouble, and someone would come immediately. When I finally went to bed that night it was quite dark, but quite peaceful in my little glen. Being one of the last to go to bed, it was very quiet as well.

Just before falling asleep, I heard an unusual clacking sound that sounded like small pieces of wood being tapped together with no particular rhythm. It sounded quite a ways away so I didn't give it much thought. There were people taking part in the retreat scattered around the house, inside and out, and a few towards the east in RVs. The sound seemed to be further east of where I was. I fell asleep and woke a while later.

I've no idea of the time as it was dark and I didn't really care to turn on a flashlight. I heard the clacking sound again, and it was much closer this time, now sounding like it was to the north of the house. It was louder and just as irregular as before, still sounding like two pieces of wood (branches?) being tapped together. I figured it was a bird rapping on a tree. I had no feeling of fear and soon fell back asleep.

Sometime later, I again woke up, and now the clacking noise was much closer to where I was, and just as irregular, sort of a "clackclack, clakkkk, clackclack clack," pause, "clackkkk, clack." The sounds were much like a woodpecker yet slower in motion, with an added hollow sound at times.

The next time I woke up there were no wood knocking

sounds; there was instead a scratching sound on the upper side of my tent. "Scratch, scratch, scratch," pause, "scratch, scratch," pause, "scratch"—again with no rhythm. My tent was a dome tent, so there's no place for a bird to land or sit to scratch. I had no feeling of fear; I felt very peaceful and felt no reason to ring the bell. I thought maybe a branch had slid down, was rubbing the tent and causing the sound, so I fell back asleep.

In the morning, I checked around my tent and found the closest thing to my tent was three feet away. Nothing around the camp was or could have touched it. I didn't give it any more thought and went to the house to help with breakfast. While in the kitchen, I mentioned all the clacking and scratching to my friend, the Leader, and a friend of the Leader standing nearby (who lived in a traditional Indian dwelling on the north side of the house). The Leader looked at me in surprise and said, "You don't want to know what it was."

Being inquisitive by nature I said, "I do want to know as I was out there and will be again tonight. Please tell me." After much back and forth bantering between the three of us, I was finally told it was a "Stee Ah Taut." I asked, "Is that was some kind of a bird?"

They laughed and said no. So I still didn't have a clue. Finally after yet more badgering by me they said that I might call it a sasquatch. That shut me up for a moment or two. I was told that the sasquatches knew I was where I was. They were aware that I wasn't afraid and wouldn't be harmful to them in any way. They were playing with me and teasing when scratching on my tent.

During the rest of the day, word spread among the group about what happened, and everyone (other than the Leader and his friend) tried to talk me out of sleeping in my little glen again. But by bedtime I couldn't wait to get to bed and listen for my new friends. However, nothing happened the second night, which was disappointing to say the least.

Since then I have talked with the Leader and learned there are several "Stee Ah Tauts" who frequent the area and have for as long as the Leader can remember. The Leader has seen them many times, as have other people in the area. Different sized ones are seen as well. Often food is left out for them, which is always gone by morning. And in return, sticks are left at the site where the food was received.

I'd love to be honored with another meeting with them. I'd be more in tune than I was then, thinking they were a nocturnal bird! Having had a greeting from these gentle souls has left a permanent impression on my life. I've cherished being "chosen" for this experience.

Elusive Creature, 1999

Submitted by: R.J.S., age 25, of The Dalles, Oregon (heating and air conditioning technician).

Incident: November 1999, east slopes of the Cascade Mountains behind Mosier, Oregon (Wasco County).

I was seventeen years old in 1999, and it was during the second half of elk season that my best friend John and I went hunting with an older friend, Frank, in his four-by-four truck. We were on the east slopes of the Cascade Mountains, about fifteen to twenty miles as the crow flies from Mt. Hood behind Mosier, Oregon.

We were heading up Husky Road, making the sharp turn onto the graveled logging road called Mosier Creek Road, and as soon as we did my buddy John said, "Look at that!" By the time I looked I didn't notice anything, but Frank pulled over right away.

We all got out and scurried over to a steep sloped bank. I saw a deep slide mark down a trail used by dirt and mountain bikes in the summer. The slide mark was quite large and it was obvious that something huge (like a large person) had slid in the mud down the trail.

There were other marks in the red clay mud that were definitely not caused by a person, and some deep depressions in the mud that I knew were unusual. There was a smell of musk or stink in the air, almost like a herd of elk or a big ol' buck in rut that time of year. John appeared to be in shock and mumbled some swear words. Frank mumbled "awesome" or something like us younger folks say. I asked, "I thought you saw an elk, what the hell was it? Why did we stop?"

John said he saw a huge reddish brown creature come running toward us on the trail that ran parallel with the road; the thing quickly slid down the bank and through the low brush. He thought it was an elk at first, then realized it wasn't, as it was running upright on two legs.

Frank said pretty much the same thing. That it was huge, red or brown, running on two feet, and he saw it go over the bank. It was like no animal they had ever seen before. They were in disbelief and it was almost unspoken between them what they think they saw.

I was really pissed that they saw a bigfoot and I didn't, especially since they were the ones who never believed in them. I suggested that we hang around a bit because I wanted to see it. It was cold and raining like it usually is in November, so Frank, being a sissy, went to the truck while John and I looked about. We were a little nervous, but I had my dad's .358 Norma Magnum and John had his .270, so I felt we would be fine.

As we walked about looking for tracks, I came across something truly unique on the ground in front of me. It was a huge, and I mean huge, pile of scat. It was definitely the biggest poop we had ever seen before or since. It was gray in color yet contained a mixture of other colored material mixed in.

It looked human, yet was not, even Shaq couldn't drop a pile this huge. I yelled to Frank in the truck, who worked up enough nerve to come over and take a look. He had the same reaction we did. I took a sandwich bag from the truck and got a sample.

In high school I had what is called a job shadow, where you get to explore possible careers for your future. I chose the Oregon Fish and Wildlife in The Dalles working with a biologist. I brought the sample on the day of my shadow and told him the story of what had occurred. He took the sample to send to the lab. I waited a long time and began to think he was just humoring me when, three long months later, he called.

It was the conclusion of the lab that it was the feces of a large omnivore and the best guess was likely a bear. There was no way in the world that I could accept this. I have seen bear droppings in the forest and these were not dropped by a black bear.

Scott, the biologist, said it was likely from a bear, but by my description of its size and color it sounded more like a grizzly, which don't live in Oregon at all. There were grasses and deer hair in the scat and what looked like pieces of pine cones (the outer part, shingle like) and some other vegetation.

Other than the slide marks down the bank, I did not find any tracks in the area, but the way the feces was still warm made me sure that what John and Frank saw go over the bank is what made the pile. I have been up to the area in the past couple of years during archery season and now pay more attention to my surroundings.

Sighting at Chinook Pass, 1999

Submitted by: J.A.H., age 39, of Richland, Washington State (fire service administration).

Sighting: April of 1999, Highway 410 along Chinook Pass, through the Wenatchee National Forest in the Cascade Mountain Range (Yakima County).

One late spring night, with my four children asleep in the car, I was driving to South Bend from the Tri-Cities and decided to take Chinook Pass, for some reason, rather than the normal Snoqualmie route. It was about 11:45 p.m. as I was driving through the forest and went past the Whistlin Jack Lodge. I am not sure of the exact location but I remember going through a few twists and turns and then coming to a straight-away where the road goes up hill. I was thinking to myself how quiet it was and there were no other cars on the road.

As I looked up the road, where the straight-away crested, I could see this figure walking across the road from the right to the left side. It was very tall, not quite up right, and walking slowly across the road. I was going about fifty miles per hour and slowed when I saw it. As I came closer I had to keep from gasping as I suddenly realized that this was not a man.

It was large and at least seven feet tall or more. It was dark in color in the night; its hair was drapy and had a long sleek look to it. If you've ever seen the show *The Addams Family,* his head reminded me of "Cousin It" in the way the shape of the head is round on top but widens out at the bottom. The legs were long as were its arms, which reached to its knees, and they swayed back and fro as it walked.

I was instantly afraid and I could feel the hair on my arms, neck, and head stand up. I kept driving and came within thirty feet of it. It did not stop and never looked in my direction, it was looking down and toward the direction it was going. It walked swiftly across the road in five long steps. There was a guardrail on the side (the metal ones with wood posts) and it stepped over this with no effort, and continued into the trees and was gone.

My hair was literally standing on end as I slowly drove past the spot it had crossed. I was scared and excited all at the same time. Had I not had my children in the car I might have been bold enough to stop and see if I could see more. I was wide-awake and the adrenalin I was feeling lasted the rest of my trip.

I drove through that area again in the summer of 2006 and recalled my memories. Once again the hair on my body rose and I knew I was in the same location. Now I know that there are lots of stories, and people see things, but I have never experienced unnatural feelings like the raising of my hair like that. I will never forget. I hold strong that I truly saw the creature and had a sighting. I am a believer.

Eyes Behind the Pines, 2000

Submitted by: J.A.M., age 20, of Mckinleyville, California (student attending college for animal conservation).

Sighting: summer 2000, Salyer Heights Road, Salyer, California (Trinity County).

One dark night, about 10:30 p.m., I was driving up Salyer Heights Road. This is about five miles outside of Willow Creek, California, in Trinity County off the Humboldt County border. The road is cut through a very secluded forested area of thick brush and trees. It's a very windy and narrow road.

As I went around one of the narrow corners, my headlights caught something moving in the trees, about ten to fifteen feet off the side of the road. I slowed down to look, thinking it might be a bear, and then realized it was moving vertically, and walking upright through the bushes. I could see the glow of its eyes. It was approximately seven and a half feet tall, judging by the bushes around it. I suddenly knew that where its head was, was too high to be a bear. And no bear could make the fast movements this thing was making.

I could see its head, shoulders, chest, and part of one arm. It had long black/brown hair that drooped over the face and down the shoulders, sort of like an orangutan. The face looked mildly similar to a monkey of some kind. It did not have a long, slender face like that of a bear, or anything else I have ever seen in the woods, especially for its size. There is no way it could have been any animal native to the area. I think the car startled and scared it off, as it disappeared into the darkness and was gone. I was too scared to stick around and find out what it was. I had my window down and quickly rolled it up as soon as I saw the thing.

About a week later, and a mile and a half away in that same area near the river, two friends of mine saw the same thing. A strong rotting smell is what caught their attention before they saw a furry, upright creature running off into the woods. Their description was of the same creature I had seen.

A Stop Alongside the Road, 2001

Submitted by: C.D.F., age 32, of Eugene, Oregon (graphic designer).

Sighting: August 2001, southwest Eugene, Oregon (Lane County).

I used to haul bundles of newspapers late at night/early morning for the local paper. One area I drove through was a fairly new road in southwest Eugene, Oregon, just east of Willow Creek Road. The area is protected wetlands near the Hynix semiconductor plant.

It was during the peak of the Perseids meteor shower that I decided to pull over because it was really dark and I could probably see the night sky better. I had to relieve myself anyway, so I got out and walked around my van to the side of the road and gazed at the sky while doing my business. I was looking up when out of the lower half of my vision I saw something move. It couldn't have been more than twenty feet away.

I saw a huge, hairy, broad-shouldered figure stand up, turn its back to me and calmly walk off into the brush. If I hadn't already just peed I would have marked my territory in my pants! I very quickly dashed around the van, hopped in, and took off.

The creature was at least seven feet tall, though it was hard to tell because where it stood was lower than the road where I was. But he was taller than me, and I am six foot one. It had no neck; its head and shoulders were mountain shaped into one angled peak. As it moved, it sort of moved its arms in a breast stoke motion to get through the brush. I could hear the branches and twigs breaking as it pushed its way through.

Later I wondered who had been more frightened, me or the bigfoot? I mean, imagine that you're the poor creature, sitting there minding your own business, maybe watching the meteor shower, when suddenly a human drives right up to where you're sitting, parks its rig, walks right up in front of you and marks his territory. If I was a bigfoot I might have been pretty intimidated too!

The area remains undeveloped as part of the protected wetlands of Eugene. I never returned to the spot, I even took another route every day after that to purposely bypass the area.

Hiding in the Bushes, 2001

Submitted by: M.E.S, age 37, of Redmond, Oregon (self-employed).

Incident: October of 2001, near Morton, Washington (Lewis County).

I am an avid hunter and was hunting black tail deer near Morton, Washington, located in East Lewis County at the foot of the Cascade Mountains, nestled between Mt. Rainier, Mt. St. Helens, and Mt. Adams. I had hiked from my parked truck about two miles into the wilderness where it was quite secluded. I found a place to sit with a good view of a clearing and settled in, hoping to find a nice buck.

After about an hour of sitting, I heard something loud rustling in the bushes on the far side of the field—about 200 yards away. I was sure it was deer or elk, based on the amount of noise it made in the brush. The rustling continued for about fifteen minutes and nothing came into the field—unusual for a deer or elk.

I could hear another sound being made that was not like a deer or elk, just soft grunts, like someone struggling with something. I was quite nervous and had my gun at ready when a large, tall, dark-brown, hairy figure stood and moved through the brush right at the edge of the clearing. I could not make out a complete body, but it was definitely not a deer or elk, and it was not a bear. I am very confident as to what those animals look like in the woods, and it was not one of those. It definitely stood on two legs and moved (walking) almost like a person through the brush. It was at least six feet tall, or better. It never entered the clearing, but I am very confident that I saw a bigfoot.

I went back to check the area the next day. I could see where the thing had been hiding, as the bushes were matted down, however the ferns and briar bushes were all very thick so I found nothing else in the dirt.

My friends tease me a bunch, but they also know that I do not lie and am dead serious about what I encountered.

High in the Cascades, 2001

Submitted by: R.S.E, age 32, of South Albany, Oregon (heavy diesel mechanic).

Sighting: November 2001, east of Albany, Oregon (Linn County).

I did see something one cold, rainy night, deep in the Cascades. I had my wife with me, and it was around 10:00 p.m. We were in my Dodge D-50 4x4 on a logging road that heads deep into the Cascade Mountains. We were probably thirty-five miles in from the main line snow peak, somewhere around R line, which is very high up; only four wheel drive vehicles can get through up there.

It was just starting to get foggy, when we saw it—a bigfoot. It was running up the road in front of us. My truck was in second gear, and when I tried to speed up to get closer for a better look, it suddenly bolted off into the trees. It was only about forty feet in front of us; it was foggy but we knew what we had just seen. This bigfoot was approximately seven feet tall and well over 300 pounds. This thing was big and it moved fast. It was running upright, its arms swinging, and it was covered in dark brown hair that was wet from the rain. It looked like a huge, giant gorilla.

Funny how it disappeared so fast. When I stopped on the road exactly at the spot where it left the road, I shut the truck off. We couldn't hear a thing, just dead silence. Yet I had the feeling it was watching us, maybe from behind a tree, just waiting for us to leave. The experience very much freaked us out and we quickly left the area and returned home.

We never told anyone until now, because we knew no one would believe us. To this day, my wife does not go with me to the mountains. I have never returned to that spot, yet I remember it so well I could take a person back to the exact spot where we saw it. This experience is forever burned into our memory. Bigfoot stays in your mind for a long period of time. This is my story, I now believe in bigfoot!

Camping Canada, 2002

Submitted by: A.M.Y., age 19, of Vancouver, B.C., Canada (university student in writing and psychology).

Sighting: July 2002, campground near Grand Forks, Canada.

When I was fifteen, my best friend Landon and I went camping with my parents for a weekend in the summer. We live in Vancouver, B.C., so we took a long drive to a campground east of Vancouver near a little town called Grand Forks.

My family arrived at the campground with everyone stiff, tired, bored, and cranky. My dad and mom checked in while Landon and I pitched the tent at the campsite where we were to stay. After several failed attempts we finally hoisted the tent. We built a fire (it was getting dark) and roasted marshmallows that night.

By the time we unpacked, lit the fire, readied the sleeping bags and got the food cooked and eaten, it was around 9:30 p.m., and the sun had just set. Landon suggested that we take a walk around the campsite to see what was there. We grabbed our flashlights and headed out.

We followed a dirt path heading deeper into the forest. We had been walking for sometime when all of a sudden we heard heavy breathing to the right of the path. We both immediately stopped, thinking it was a bear (we each lived by a big park where bears had been spotted several times, so the two of us were extremely paranoid). We quickly stepped way back putting distance between ourselves and the creature and shone our flashlights into the dark bushes and trees.

What we saw would question the work of most scientists to date about the complex web of the animal kingdom; if they saw what we saw, they would most definitely add the creature "bigfoot" or "sasquatch" to the long list of primates.

The creature that we first thought was a bear had been crouching or kneeling, and as we shined our lights on the bushes it grunted and began to slowly rise to stand on two legs. It kept getting taller and taller behind the bushes. I remember thinking, "When is this creature going to stop getting taller?" When it was fully erect on two legs it stood about eight feet tall, I know this because Landon and I were close to six feet and this thing was more than two feet taller than us. It was covered in dark brown hair and had the posture and body structure of a human.

With our mouths open, the two of us could not move; we could only stare while this huge creature, through the dark hair covering most of its face, stared back at us. Then it turned and ran swiftly into the dark forest. We were scared out of our minds and too afraid to follow it.

We headed straight back to the campsite to tell my parents everything that happened, but they didn't believe us. We were so terrified neither of us slept a wink that night, although we secretly hoped it would come back, if only so my parents would believe our story. Landon and I still talk about that night; we know what we saw.

The Creeper, 2004

Submitted by: Shane Lanes, age 35, of Eureka, California (grand adventurer).

Sighting: March 2004, Dunsmuir, California, south of Mt. Shasta (Siskiyou County).

I don't tell this story much, but in 2004, my friends, Roofie Roof Top, Sparkles, Scott, and I, hoboed our way to Dunsmuir, California. We had been at a campsite just shy of a week, and were having a great time sitting by the campfires, eating good food, and just enjoying each other's company. The next morning we were going to hop a train south to Roseville, and then a short jaunt home to Sac Town (Sacramento).

My friends rolled out their sleeping bags around midnight and quickly fell asleep. I was real excited to be leaving for home the next day, so I was up most of the night with the gears in my head turning, thinking about the things I had planned to do when I got home.

I lay there for hours, listening to the roar of the river and looking up at the beautiful full moon when all of a sudden, between my feet and brother moon, walked an extremely tall, slightly hunched over dark figure that was not human!

It was hard to be precise of its height since I was lying down, but I can honestly say it had to be upwards of seven feet tall, or more. A dark brown was the color of its thick, hairy body. With eyes wide open, I quietly watched as this great mystery walked smoothly past without making a single twig snap.

I was quite impressed with its stealth. Without a care, and with giant strides, it headed down the ravine to my right. To this day, I can still feel my heart rush from the memory.

To the defense of my friend, the bigfoot, I was freaked out, yes, but I did not feel threatened in any way. When I later told my friends of our campsite visitor, they had mixed feelings of doubt and amazement. Mostly doubt, of which I have none. I know what I saw.

A Night on the Skykomish, 2004

Submitted by: K.L.M., age 41, of Everett, Washington (private investigator).

Encounter: spring 2004, Skykomish River, Washington (Snohomish County).

We always camp in the same area, twice, maybe three times a year, all the way up Index Road, second from the last non-official campsite and right on the Skykomish River. There is an embankment leading down to a small beach patch at the river's edge; it's only about forty yards to the other side, where the river narrows a bit. The other side is full of wildlife; bears and bobcats are seen all the time. We have spotted bears catching fish in the spring when the river is raging at its height of runoff.

My wife and I have a tradition of hiking to the top of the river at least once a year; this particular time we strayed too far from camp, got caught up river and had to hunker down. When the river is raging as it was, it's barely negotiable during the day, much less trying to get around it at night. It was pitch black out, very little moonlight as it was overcast. We found a hollow tree and hoped it didn't belong to a bear. It gets cold up there at night.

While sitting next to each other huddling for warmth, we heard loud footsteps coming closer toward us from in the river. We could not imagine what it could be, as it was much too noisy and loud for a human, and bears don't stomp on two feet. We knew it was bipedal from the sound; the water in this particular part of the river ranged from four feet deep to very shallow closer to the bank. We heard grunting, a lot of grunting, like it was struggling through the slippery rocks. The stench coming from the creature was strong and putrid, almost unbearable, and we became frozen with fear.

In the dark your senses are very alert; you can tell when something is moving around you. We heard it sit down. It was huge, not like any bear we have ever heard in our camp while in a tent. It actually sat on the bank about thirty feet from where we were hiding. My wife was shivering so hard that it must have either heard or smelled us. It instantly went dead silent. We knew that it knew we were there.

All of a sudden, we heard sniffing right over our heads! My wife screamed and so did the creature. Blood curdling! So loud I thought my ears were bleeding. The scream was piercing, lasting about forty-five seconds. There were a few hoots after, like a giant gorilla would do. Then it bolted across the riverbed and up the side of the mountain, very fast and very furious. I looked out and saw a huge, dark, shadowy figure on two feet disappear into the woods.

We were scared frozen for quite some time, perhaps twenty minutes or so when we decided it might come back. At that point we ran back to our campsite, packed up as fast as we could, and sped off down the mountain.

I did call the sheriff when I was back within cell phone range, but there is only one officer on duty for the entire county on that side of Interstate-5. My wife refuses to even hike anymore and we have not been back. That smell is still with me, I will never forget it.

Truck Driver Joe on I-90, 2004

Submitted by: H.A.D. (for and about his friend Joe), age 52, of Auburn, Washington (truck driver).

Sighting: early summer of 2004, Interstate-90 highway, Snoqualmie Pass, Washington (Kittitas County).

In the early summer of 2004, forty-eight year old Joe, was traveling east to Spokane from Seattle in an eighteen-wheeler. It was about four in the morning as he was driving up Snoqualmie Pass (Highway I-90). He was driving between exits forty-eight and fifty-one, and his truck was moving slow as he was climbing a hill.

As Joe was driving slowly along, he saw a strange hairy

creature, about eight feet tall with long arms, standing on the right side of the road. It reminded Joe of a large ape, with an egg-shaped head.

As Joe was about to pass the "ape creature" he began looking out his passenger side mirror, when suddenly the creature began running along side the truck, as if it was trying to keep up with him.

The creature got closer and closer to the passenger side door, its dark brown, mangled hair blowing back from its face and body as it ran. At one point when reaching the passenger side window, it turned its head to the left and looked directly at Joe.

Joe saw one of its eyes and described the color as darker than dark. It was the scariest thing he had ever seen in his life. It frightened him so much he said, "I almost messed my pants right there." After a few minutes, the bigfoot swerved to the right, leaped over the road barrier and was gone.

Joe was so terrified that he never stopped at the town of George where he was supposed to do his log book, and instead drove nonstop to Spokane.

Hunting Mt. Ranier, 2004

Submitted by: C.A.B., age 50, of Grand Coulee, Washington (firefighter/EMT, owns a scuba diving business).

Incident: Fall 2004, Mt. Rainier, Washington (Pierce County).

I am a volunteer firefighter and EMT, and have been hunting since I was nine years old. A couple of years ago, my hunting partner, Ace, and I went hunting at the base of Mt. Rainier around the 7,000-foot level. At the bottom of a dark canyon surrounded by old-growth timber, we hit a split in the trail, and decided he'd go left and I'd go right.

I made my way for a half-hour or so when I noticed an odd odor. I tried to figure out what it was; I have spent more time in the woods than most and had never smelled anything like this before. It seemed a cross between mold and wet hair.

I started back up the trail and noticed the smell getting stronger. I got my gun ready, as the only thing I could think of was maybe it was an old bear and he had rolled in something.

I then noticed that about three feet off to the right, something had been walking alongside the trail, leaving tracks in the brush. No prints really, just smashed down vegetation and such. After looking it over I realized that the steps were several feet apart.

That's when I started to get a little nervous, as bears don't walk upright like this. I moved up the trail a little further, telling myself it's time to turn around and get the heck out of there, but something (curiosity? stupidity?) made me go on.

A few feet further and I stepped on a limb that broke with a loud "crack." That's when all hell broke loose. I heard loud crashing through the brush and tree limbs snapping; a tree cracked loudly and hit the ground, followed by loud splashing sounds. And then dead silence for several seconds. Followed again by more loud brush crashing and limb breaking sounds. I just about messed myself. After a few minutes, I decided that whatever it was had left and I slowly proceeded up the trail, shaking all the way.

As I came around a small bend I found something that made me a believer. There was an old-growth cedar stump about twenty-five feet tall, and a large portion of it had been broken off around eleven or twelve feet up. At the base was a partial footprint that slid from the base of the stump into a huge mud puddle. There was muddy water dripping from everywhere.

Something had obviously tried to go past the stump and slipped when it stepped on the slick roots and fell into the water. As it fell, it grabbed onto the stump to catch itself and tore part of the stump away. And as I said, the missing piece was eleven or twelve feet up.

I never saw anything that day, but I know from the smell and the evidence I found, that a bear would never have slipped like that and they definitely do not walk upright for long distance. I still hunt that area and hope that someday I can actually see what I missed that day.

The Awakening, 2004

Submitted by: S.D.W., age 32, of Marysville, Washington (home-schooler/in-home child-care provider).

Sighting: November 2004, Silverdale, Washington (Kitsap County).

I've heard that some people wait their whole lives to have an experience like I did. I have seen this creature. I know there are those reading this who would scoff and make light of my experience. Believe me, there are days that I don't really like to dwell on it. And other than immediately after, I never talk to others about it.

That cloudy, dewy morning, I ended up calling 911 and crying to my mother-in-law about how terrified I was. My reality was changed. My sense of false security had changed and my belief in things concrete, things a hundred percent factual, had vanished.

My husband, small children, and I were living near my mother-in-law in her basement apartment. I was working graveyard shifts in Bainbridge Island and coming home around 7:00 a.m. This particular morning, the fog that hung over the road was dense, but I could see well enough to drive. I was driving up Anderson Hill Road in Silverdale. I traveled alongside the huge power lines and looked around as always, taking in the area.

Around the area is a soccer field, a home on a few acres, and lots of woods. There is a ravine that dips down and gradually comes to level with the road just to recede back down below eyesight. It was this area that I saw something moving from my right peripheral vision. I wasn't going very fast because deer visit this area and run across often. I instantly looked toward this moving object.

It was huge! Maybe eight or nine feet tall. It was hairy and looked much like the massive body of a large ape, but it had no neck. The color was like mahogany and it had eyes like a human, but I am not sure how I would know this, because my conscious memory seems to remember no eye contact at all. He did turn his head slightly to the left toward the road, but no eye contact was made.

The creature was walking with his arms beside him and was totally oblivious to my mini-van. He was not looking at me or the road, and was walking quite peacefully, yet with a purposeful intensity. He had a destination. I felt like he was intelligent, or at least knew where he was going as he walked along the power lines.

I also noticed that it was hunched slightly like a dowager's hump or fat pad on his upper back. His arms seemed longer than they should have been for a man, so I acquainted him more with an ape. His arms hung below his waist but no knees were visible. He was so hairy and massive. "What the heck is that?" I remember looking away to see if anyone else was around, and when I looked back over, he was gone.

"Did I just see what my eyes saw? No. Yes, I did." Almost instantly I panicked and started to drive a little faster, no longer looking at the side of the road. "I need to get home, I need to call the police. Someone needs to get out here and get whatever that was out of our woods."

I got out of my car and slammed the door. I started my car alarm thinking it would frighten animals away. I ran inside and locked the door. I called 911 immediately. The operator said, "Calm down and the first thing I want you to know is that I believe you." Just her saying this calmed me down and I proceeded to tell her what and where I saw this large animal/creature/man.

After that call, I called my mother-in-law upstairs and I am sure she believed me, or at least believed that I believed what I saw. She told me that there was a boy from our church who saw the same thing along some power lines in Silverdale a few years earlier, and encouraged me to call on his family and tell them. I did this, but not until Christmas time, about one month later.

I have rarely spoken of my bigfoot sighting to anyone. On a side note, I get the feeling from this experience that my mind was awakened to the impossible. I now believe anything is possible. I know what I saw. I now believe what others say they saw.

And I will say this; it is sometimes like a dream when I recall it. Was I in this dimension of life or did I peak into a different world? Why did this thing not see me? Where was he going? To this day, I hate driving up Anderson Hill Road. And every single time I pass this area I keep my eyes fixed on the road. I don't want to see it again.

The Forest Park Sighting, 2005

Submitted by: Gerald Edmund Lunn, age 20, of Portland, Oregon (student and audio/visual technician).

Sighting: mid-January 2005, Forest Park, northeast of Portland, Oregon (Multnomah County).

On a cold, dark night in January, my girlfriend and I went to Forest Park, which sits just northeast of Portland, Oregon. We parked in a tiny lot off to the side of the road surrounded by forest. It was around eleven o'clock at night, and other than the moonlight and a streetlight, it was very dark. It had been lightly drizzling off and on, there was a breeze and the trees were swaying. Before long, I began to hear noises, bushes moving, rustling, and twigs snapping. It never got very loud, but it was definitely noticeable. At first I thought it was the wind.

Suddenly, I began to hear definite movement and footsteps coming from out of the woods, the noise becoming louder and louder as if someone was approaching, which was alarming, as we were the only ones parked in the dark lot. I figured that maybe it was a park ranger coming to tell us to leave. I couldn't see anything from inside the car, so I stepped outside. And there, approximately fifteen feet away from me stood a bigfoot creature!

He seemed to be about six feet tall, tall but not unnaturally so. He had long, dark hair, probably brown, but it was dark so it was hard to tell exactly. I did not get a good look at the face or eyes, and I didn't smell anything. He was just walking at a very casual pace, and turned from us to go back into the woods.

Was he looking in the car? Was he watching us? I wasn't afraid when I realized what was happening because he was already walking away and because I've never heard of a bigfoot attacking anyone. My girlfriend was a little freaked out, just by the fact of seeing a bigfoot.

I've told friends about this, but for the most part they're skeptical. It doesn't matter to me if people believe or not because I'm sure people will accept it in time.

Note: In 1803, William Clark (of Lewis and Clark fame) paddled far enough up the Willamette River to see Forest Park's present location. He described this forest as having Douglas fir as its predominate tree, with trunks ranging from five to eight feet in diameter. An abundance of wildlife can be found in Forest Park. The park contains over 5,100 wooded acres, making it the largest, forested natural area within city limits in the United States.

Trinity Alps Watcher, 2005

Submitted by: bigfoot researcher Sean Fries of Weaverville, California (U.S. Navy combat medic for six years; served in the first Gulf War).

Sighting: June 20, 2005. Trinity Alps wilderness, near Parkers Meadow, California (Trinity County).

I was backpacking with my ex-girlfriend Susan, investigating two reported bigfoot sightings in the past three weeks in the area of Trinity Alps Wilderness. One was from an

MD who was fly-fishing. We made camp for the night up along Swift Creek Trail and turned in.

It was very quiet all night long. We got up around 7:00 a.m. and Susan began cooking some breakfast on the pack stove. I was busy getting a fire started, as it was really cold that morning.

We had brought along one of those "Wheels and Deals" free newspapers to start a fire with. Susy had it tucked into the mess straps on her backpack. During the hike, it fell out somewhere, therefore that night we couldn't use it to start a fire, so we ended up using an emergency flare to get a fire started.

Well, the next morning, as I was gathering some kindling for another fire, on the far edge of our campsite I found a little bed of pine branches laying on the ground, with that newspaper laying right in the middle of them.

Susan later told me while cooking that she had gotten a feeling that she was being watched.

There was a creek between us and the tree line, called Parker Creek. She turned around to look directly behind her and was the first to see a creature standing in the tree line watching us.

She was very scared and quickly spun back to me with a shocked look on her face. I had seen the movement of her spinning back towards me and looked up at her to see a look of utter disbelief.

I looked in the direction of where she had been looking and standing in the tree line was a creature about seven feet tall watching us. I stood up and walked toward it to try and figure out if it was a bear or what, as it was partially blocked by the trees.

As I got around Susy, the thing's whole upper body turned away from me and it walked on two legs off into the trees. I quickly pulled my boots on, grabbed my camera and gun and took off after it, but it was gone.

The creature had dark brown hair maybe two inches long and very broad shoulders. I really focused on the eyes. It had yellow eyes, and a broad nose.

When we left, I left some bacon on a log at the campground. We came back three weeks later on a day hike to look around. When we got to the area where the incident occurred, we began to hear the sound of two rocks being hit together. This went on for approximately twenty minutes. I think it remembered us and was letting us know it was still around.

Eerie Feelings, 2005

Submitted by: T.C., age 42, of Darrington, Washington, Snohomish County (transportation worker, D.O.T.).

Incident: fall 2005, southeast of Darrington, Washington (Whatcom County).

It happened in the fall of 2005, when the leaves are still falling and the campgrounds are closing for the season. My girlfriend and I were hiking down a closed dirt road that leads to a camping site southeast of Darrington. The weather was cool and cloudy but it wasn't raining; a nice day for a walk.

We had only hiked in far enough to be out of sight of our car when it suddenly became eerily quiet. There were no birds, squirrels, or anything moving. So quiet you could hear the leaves fall from the trees. I heard one leaf a good ten feet away tumble through a tree to the ground. Too quiet, unnaturally quiet.

We had our dog with us, who is normally quite fearless. She became very nervous and kept stopping and looking back toward the car; we had to really coax her to come. About half a mile in, she stopped and absolutely refused to go any further. The silence was intense and we had the overwhelming feeling we were being watched.

With the eerie feelings and the dog refusing to go on, we decided to turn around and head back to the car. About halfway back, we came across a huge pile of scat in the road that wasn't there when we came in. There was no evidence that an animal had scratched around or tried to bury it. It looked human, only gray in color. We left the area and have not had a chance to go back.

Note: Worthy of note is the strong overwhelming feeling of being watched that is sometimes associated with being in the presence of a bigfoot. Not just a mild feeling, but an overwhelming feeling. Before the creature is ever seen, if seen at all, this feeling permeates the senses with the understanding that you are not alone.

Field Trip Sighting, 2006

Submitted by: Julio Barrera Jr., age 13, of Plumas Lake, California (sixth grade student).

Sighting: summer 2006, Sly Park in the El Dorado County forest, California.

Two years ago, when I was eleven years old, I went on a camping field trip in the El Dorado County forest. That night, when I was supposed to be asleep, I was looking out the window of our cabin.

About forty to fifty feet away, I saw a huge, hairy creature standing behind two oak trees. It stood like a human and looked like it was about six or seven feet tall. It kept bending over picking something up; I couldn't tell what it was because it was dark outside.

I ran and tried to tell the cabin leaders but they wouldn't listen to me; they just told me to go back to sleep. I went back to look again and the creature stepped back into the forest. It looked back once and then ran away on two feet.

The face looked kind of like a man mixed with a monkey or gorilla. The hair looked black from the dark, but I knew it might have been a darkish brown color.

The next morning, my friends and I went outside to check it out, and the area and the trees smelled real nasty like wet dog with a sour food smell, which almost made me puke.

Multiple Mysteries, 2006

Submitted by: T.E.C., age 43, of Lake Cushman, Washington (retired cashier).

Sighting: summer 2006, Lake Kokanee and Lake Cushman area, gateway to the Olympic Mountains, Washington (Mason County).

> I live in the Lake Cushman area above Hoodsport in the Olympic Mountains. I have had several incidents and two sightings. Last summer, my friend and I were fishing in a boat in Lake Kokanee. We heard something big running in the water. Big enough to create waves. We could hear how the water was splashing and it was obvious something big was running through it. We were about thirty feet from the shore and the waves did rock the boat. We noticed a really bad smell, almost bad enough to make our eyes water.
>
> We also noticed that the birds and the whole area got very quiet and stayed that way. There is an inlet that goes through the woods. This thing must have stepped out of the lake and ran into the woods and everything got quiet. My friend decided we should head back. As we were leaving, I looked back toward the inlet and the woods and noticed something very big and dark standing by the trees.
>
> It was standing on two feet and swaying back and forth. It was in the shadows, but I could tell it was hairy, had long arms and stood maybe eight feet tall. I felt like it was looking into my eyes. I told my friend to turn around and look, just as the thing stepped back into the trees.
>
> The second sighting was very brief, and the creature looked the same. This took place last year as well. I was driving home at around 11:00 p.m. In my headlights, I saw this huge thing run across the road in front of me, jump a deep ditch and disappear into the woods. It was running on two feet, its arms swinging at its sides, dark and hairy, and again about eight feet tall.
>
> The thing is, I never before saw anything run that fast. And it never looked at the car at all. Which I thought

strange because any animal caught in your headlights will usually turn to look at you. I went back the next day to have a look at where it ran into the woods, but couldn't find anything as we had a big windstorm that night.

I live above Lake Kokanee. I hear these roars between one and three o'clock in the morning. It is like nothing I have heard before. As strange as it seems, it almost sounds like that furry thing from *Star Wars*. Other people have heard things around here as well. Another friend who I go fishing with hears the roars from over at his place too.

I have come across curious rock piles up in the mountains. I at first just thought someone was bored and piled rocks. Several times late at night, I also hear these wood thumps. I've been camping further up the mountains and I've heard these same thumps as well. At home, late one night, I was thumping on the trees and got several answers from different directions.

Then something heavy came crashing down in the yard on the other side of the house. I ran inside before I saw anything, but I felt like something was watching me from the outside. And my cats that were sitting in the windowsills were all freaking out and afraid of something. I have lived here for about five years, and it has been since the Bear Gulch fire that all this really started.

I plan on more exploring in the mountains as I now have a four-wheel drive vehicle. Bigfoot makes me curious enough to do this. I want to know what it is and I believe they mean no harm. I hope one day for a closer encounter.

Chapter 4

Interviews – Researchers

Ignorance more frequently begets confidence than does knowledge: it is those who know little, and not those who know much, who so positively assert that this or that problem will never be solved by science.
—Charles Darwin

DR. JEFFREY MELDRUM
October 2007

Dr. Jeffrey Meldrum

LCS: Hello Dr. Meldrum. Thanks very much for agreeing to do this interview. For the few people out there interested in cryptozoology who don't know who you are, could you tell us a bit about yourself?

JM: I am a professor of anatomy & anthropology at Idaho State University and affiliate curator at the Idaho Museum of Natural History. My formal study of primates began with doctoral research on terrestrial adaptations in African primates, and has since taken me from the dusty, skeletal cabinets of far-flung museums, to the remote badlands of Colombia and Argentina in search of fossil New World primates. I have published extensively on the evolutionary history of the South American primates and have described several new, extinct species.

I have documented varied primate locomotor specializations in laboratory and semi-natural settings. More recently, my attention has returned to the emergence of modern human bipedalism. I've co-edited and contributed to, *From Biped to Strider: the Emergence of Modern Human Walking Running and Resource Transport* (Kluwer–Plenum, 2004). My interests also encompass the evaluation of footprints, both in the lab and in the field, attributed to sasquatch. My book *Sasquatch: Legend Meets Science,* an expanded companion volume to the Discovery Channel documentary, was published by Forge/Tom Doherty Associates in 2006.

LCS: How did you first come to be interested in cryptozoology, and specifically bigfoot research?

JM: Many familiar species began as cryptids, exotic animals often embellished by local legend and mystery. I think all zoologists have a certain fascination for the discovery of new or elusive species of animal. Growing up in the Pacific Northwest, I was familiar with the legend of sasquatch or bigfoot, which fell in step with my interests in primates and prehistoric humans. But it wasn't until I had the opportunity to examine fresh tracks myself, that the hook was set. I had to get to the bottom of this.

LCS: In your opinion, what is the best piece of evidence for the existence of these animals? Why are you so convinced that they are real?

JM: Obviously, my attention was drawn to the footprint evidence. It is the most ubiquitous "tangible" evidence at hand. From my position of experience and expertise it seemed that was where I could make a meaningful contribution. I have assembled a large sample of footprints in my lab and from the study of these specimens I have inferred a model of the sasquatch foot that is both distinct, while internally consistent, and appropriate to the observed locomotion and associated habitat texture. We are now creating an archive of 3-D virtualizations of these casts to facilitate dissemination and analysis of these data.

LCS: There have been several television documentaries about-

bigfoot, many of which you have appeared in. One in particular, *Best Evidence,* seemed to be anything but. The program seemed entirely slanted toward the skeptical viewpoint, and didn't allow you or any other bigfoot believers to fully make your case. Do you agree with this assessment, and what would you like to see changed in the way these documentaries are done?

JM: Most documentaries are constrained to present a "balanced" treatment, which means they have me and/or another colleague, such as Dr. John Bindernagel (as in the case of *Best Evidence*) comment briefly in a few sentences, summarizing years of research and experience in the field, then bring on a skeptical scientist, or merely an armchair skeptic with little or no expertise, who may be seeing the evidence for the first time, to offer a counterpoint, which of course appears to be given equal weight. That can be frustrating and misleading. This was one of the principal motivations for writing the book, *Sasquatch: Legend Meets Science.* It gave me the opportunity to provide greater breadth and depth to the discussion, while also responding to some of the misstatements of the detractors.

LCS: Many eyewitness reports describe a strong, even overwhelming feeling of being watched. Why do you suppose so many people pick up on this?

JM: This is a perception that seems to be very real whether one is being watched by a predator or a person across a room. Obviously, the physical or psychological mechanism for such a perception is not fully understood. One possibility that has been proposed is the use of infrasonic vocalizations to elicit unease, fear, and disorientation in the witness who is perceived as a threat. Infrasound is employed by a wide variety of organisms. Great apes have outpouchings of the larynx that could be utilized to generate such low frequency vocalizations. Perhaps sasquatch uses this strategy to ward off human contact. We can only speculate at this point.

LCS: Can you talk about the foul odors commonly described in conjunction with bigfoot sightings? Are they just unclean animals? Is it a pheromone released only when they feel threatened?

JM: Actually a foul odor is only infrequently reported, suggesting that it is not the usual fragrance of a sasquatch. Again drawing precedent from the great apes, male apes have well-developed axillary glands in their armpits and emit a strong pungent odor when agitated, just as humans do to a lesser degree while taking great pains to mask it.

LCS: Probably the most abundant evidence we have is in the form of footprints. Their feet seem to share traits with humans and other primate species. What is the significance of the footprint evidence in the big picture?

JM: There emerges a consistent pattern of anatomy that indicates the retention of primitive traits of a flat flexible foot, lacking the uniquely human characteristic of the arch of the foot. This combination of a non-divergent big toe and a flexible midfoot appears to have precedent in the early hominid fossil record. I suspect this foot morphology evolved in parallel in a large terrestrial ape in East Asia, and therefore provides a fascinating contrast to the evolution of human bipedalism in Africa.

LCS: Are there any plans in place in the event of the capture of a bigfoot? Where would it be taken, what would be done with it? What could we do to ensure the protection of the species and their habitat?

JM: There have been numerous discussions, as you might imagine, and some real possibilities explored, but no formal arrangements made. Obviously, documentation and verification would be high priorities. I am sure there would be a cascade of events set in motion that we can only imagine and some that I probably haven't even thought of yet. In reality however, the objective of a capture is not realistic at this point (assuming they indeed exist). Our principal objective is acquiring a DNA sample.

LCS: Are there any theoretical techniques you have in mind that might improve our chances of finding that definitive evidence we need? Anything untried, but that you think might work?

JM: As I mentioned, it would seem that the collection of tissue for DNA analysis would be the number one priority, short of obtaining a whole specimen. My colleague, John Mionczynski, is a professional wildlife consultant with experience using hair snags in grizzly studies. Employing modifications of this approach would seem to be a reasonable next step.

LCS: There are so many questions about this animal that remain unanswered. What is the most burning question in your mind that you want answered?

JM: Obviously, the primary question of their existence. Assuming that for the moment, the next question is their nature, most fundamentally how ape-like or how human-like they are.

LCS: In closing, what do you have to say to the cynics and skeptics of the world? Any final thoughts?

JM: Skepticism is a single thread in the tapestry of scientific exploration and discovery (cynicism has no place). Those who have become fixated on that single element and tug on it alone, spoil the fabric and miss out on the marvel of life's experiences.

LCS: Thank you very much, Dr. Meldrum. It's truly an honor. Keep doing what you do.

Dr. Meldrum's book, *Sasquatch: Legend Meets Science* (A Tom Doherty Associates Books, New York) was published in 2006. It provides intriguing information on all aspects of the sasquatch/bigfoot issue with numerous photographs and illustrations.

DR. W. HENNER FAHRENBACH
March 2007

Dr. W. Henner Fahrenbach

LCS: Hello, Doctor Fahrenbach. Can you tell our readers a bit about yourself and how you came to be interested in this kind of phenomena?

HF: Backpacking in most of the Western mountain ranges from the 50s onward by itself gave me an appreciation for the amount of wild country that is available to any creature that wants to hide. I was also aware of scattered reports prior to the publication of the Patterson movie. Lastly, my place of work constituted a logical place to come to with purported sasquatch hair, material that inevitably ended up in my hands.

By this connection, I eventually had the opportunity to converse with eyewitnesses, to assess their credibility and begin to acquire some insight into the habits and whereabouts of the sasquatch. That introduction then gradually led into active investigation of recent sightings, interviews and some specific, sasquatch-directed outings, on which I have repeatedly heard their screams, calls and roars, found footprints and actual track-ways, and got close enough to smell the animal.

LCS: Who does more damage in discrediting the study of bigfoot, the hoaxers or those who connect bigfoot with UFOs and mystical powers?

HF: I am adamantly opposed to any paranormal interpretation of initially inexplicable or misunderstood phenomena, the method of thinking usually being based on ignorance and lack of critical thinking. It is vastly more likely that the mind is playing tricks on the unwary rather than that the laws of nature get suspended in some trivial context. Pairing two exceedingly unlikely events in any case stretches credibility beyond reasonable limits.

Hoaxers have always been easy to expose by investigators with knowledge of the subject, except that uncritical reporters appear to relish such stories out of proportion to their credibility and rarely seem to bother to inform themselves of the fact. For example, the often quoted "fuzzy images" of the Patterson sasquatch are clearly based on casual watching of the movie rather than on a study of the dozen or so remarkably crisp prints that have been made from various frames.

LCS: What would be your ideal scenario to prove the existence of these animals? A live capture? A dead body found? Crystal clear video evidence? What's the most likely way for it to happen?

HF: In this day of computer animation, it is unlikely that even the best film is going to provide ultimate proof to skeptics. Even a fresh corpse would require analysis that very few laboratories would be prepared to offer, nor would be inclined to embark upon, given the uproar it would cause with the general public and the press. One can hope that an accidentally killed sasquatch, perhaps one run over by a large truck, would end up in the hands of reputable primate anatomists rather than being quietly buried or kept for material gain.

LCS: What's the theory of why there's never been any bones recovered?"

HF: There are, from what I have heard, bona fide bear researchers who have never found a dead bear in the wilderness that had died of natural causes. For the sasquatch, that is much more unlikely for several reasons. First, there are probably between 50 to 100 bears per sasquatch in North America— right there the odds decrease dramatically. Secondly, any sasquatch that was either ill or had been injured would try to find the deepest concealment possible, probably far deeper than they would normally use to hide in.

If the creature died in such a location and was not guarded or attended to by others of its kind, its body would be shortly taken apart by bears and coyotes, bones by assorted rodents, hair by moths and the more solid bones would be exposed to the fairly acidic soil that prevails in our forests. Such acidity would leach out the mineral content of the bones and allow the rest to be eaten by most animals. It being a deep hiding place, the first leaf fall would cover over the remains of the carcass and after one season of rain, snow cover and melting, nothing would remain visible from the surface.

After all, there used to be millions of buffalo roaming the western plains to the point where somebody calculated that the plains ought to be covered several feet deep with buffalo bones if they didn't rot and get degraded by natural processes. And all that is disregarding the remote possibility that sasquatches do something to conceal a corpse of one of their own by internment, something that we have no evidence for other than one report of a baby sasquatch having been buried.

Assuming that *Gigantopithecus* lived in the same environment as the sasquatch, then the dearth of its fossil remains, mostly two jaws and many isolated teeth, would attest to the unlikelihood of fossilization in this forested environment.

LCS: Different witnesses report slightly different physical appearances of these creatures. Are there different breeds or races of bigfoot out there? Perhaps different cultures?

HF: Just look at ten people you meet on the street. No two look alike in height, body build, weight or proportions. The same would apply to the sasquatch, probably more so since it is a small population of animals whose survival after birth is not dependent on avoiding predators but only a function of natural infant mortality. So virtually all normal variants in appearance would be likely to survive.

Given the distribution of the sasquatch all over North America and the time frame over which they have most likely been here, it is not improbable that they have developed divergent traits just like you see in the North American native population. That wouldn't be called "different breeds," and even "race" implies a more distant relationship. And all of that has nothing to do with culture, something the sasquatch has almost nothing to show for.

LCS: Certain audio recordings have been made of "bigfoot language." It sounds almost like gibberish. Do you feel these are the sounds made by actual sasquatch? And if so, could this be an actual language?

HF: You are presumably referring to the Sierra recordings. I am quite sure they are authentic since I have talked at length with several of the people involved in that situation, including two of the people who made separate recordings. There has also been a sonogram analysis done of them that you should be aware of. I am dubious that these sounds represent what could be called a language—it does not seem to be conveying specific information, as far as we know—but it may be some sort of primitive precursor to actual information content communication by vocalization. Screaming and simple beating on trees would not fit that category.

LCS: Is there any evidence that suggests whether bigfoot are vegetarian, carnivorous, or omnivorous? What do you suspect makes up their diet?

HF: That you can quite easily answer by reading a lot of eyewitness reports. These will refer to the sasquatch eating just about anything that has nutritional value—i.e., it is an omnivore—though many of these items are not on the human menu. However, under times of duress, the Native Americans ate a surprising spectrum of animals and plants out of their environment.

LCS: Seeing as they are so few in numbers, do you think this is a sign that they are diminishing, becoming endangered? Could it be that bigfoot is now on the verge of extinction?

HF: No, I don't think that the sasquatches are endangered, particularly since one of their staples—deer and elk—have increased tremendously in numbers since the white settlers arrived in North America. Besides that, both agriculture and the aftermath of clear cuts provide lots more browse than old-growth forest ever did. I also don't think reports have decreased, especially since I have been hearing sasquatch vocalizations in pretty close proximity to civilization right here. I am sure the creature will and does adapt to the higher density of people and our terrain offers no problem to thorough hiding at all times.

LCS: What are your thoughts on the migration habits of sasquatch? Do they migrate? Is it possible they might travel because of weather, the winter cold, food sources, the pursuit of mates?

HF: I am not aware of any compelling reason or evidence why bigfoot should migrate at all. They are, after all, seen in Alaska, sometimes during wintery weather seasons. I don't think bigfoot in the western states would have any reason to migrate, other than perhaps to a lower altitude during the winter.

The sighting frequency is most likely more a condition of the abundance of people to see them, rather than that the sasquatch only went there at those times. Much of the Oregon high country is inaccessible until late in the spring because of snow. Snowshoers every so often encounter footprints in the snow, although I would suppose the sasquatch would find better conditions a bit lower, though a lot wetter and uncomfortable. But then it snows sometimes at sea level and you don't see a sudden exodus of animals (or sasquatches). And it has been suggested by some researchers that there exist a cline in height and color from south to north, the taller and lighter ones living farther north. For such a gradient to exist or to become established, the respective populations would have to be very much resident in those regions and definitely not migrate. I have no opinion on this particular subject and believe that the numbers probably don't allow us to draw any comfortable conclusions. Since the principal winter food source would be deer or elk, they stay around all year and don't migrate other than a bit in elevation.

To provide a true assessment of the density of sasquatches in a given area throughout the year, you would have to find out the density of people there to observe bigfoot and factor that into the equation. I have done that at least for day vs. night and found that they are much more often seen at night than if their daytime elusiveness prevailed during the dark, meaning that they are out and about more during the dark than the people who observe them.

LCS: Were many bigfoot lost in 1980, when Mt. St. Helens erupted?

HF: I don't think that many bigfoot perished in the Mt. St. Helens eruption, since, in any case, they tend to be widely spaced—I have on occasion estimated their "territory" on the basis of repeated sightings of an identifiable animal and have come up with several hundred square miles per male or his little group, so its not improbable that they might have given the mountain a wide berth once it became active.

LCS: What do you say to all the skeptics and cynics out there?

HF: Ask them if they are familiar with the evidence (they won't be). If they say, "It's all faked," you can point out that there are hundreds, if not thousands of footprints of all sizes, far removed from any supposed hoaxers, found in remote and inaccessible areas, showing movement of the foot from step to step in keeping with the ground, and above all, a step length that exceeds anything a person can hoax. The step length, incidentally, is generally ignored when talking about fake feet and is very difficult to hoax! Remember, even an "average" sasquatch has a step length for calm walking of five feet (you can find all that in my graphs). The longest step length for which I have highly reliable measurements is thirty feet, or three steps covering ninety feet. The longest human triple jump—three steps—in essence is sixty feet, twenty feet per step.

LCS: Thank you very much for your time. It's been a pleasure.

HF: Same to you. Good luck!

ROBERT (BOB) GIMLIN
April 2007

Bob Gimlin

LCS: Thank you very much for consenting to this interview Mr. Gimlin. It is an honor to have one of the most famous names in the bigfoot phenomenon participating in this book. For anyone who doesn't know, can you tell us a bit about yourself and how you became involved in this research?

BG: Well, mainly how I became involved is because I knew Roger Patterson as a friend, and he kept wanting to go with me when I rode horses in the mountains. And he'd play tapes of people that I didn't know, but it'd sound like reputable people who'd seen footprints and had footprints. He showed me plaster casts of them, and so that's mainly how I got started and interested in it. Of course I was a skeptic, to an extent, all along. I didn't really understand a lot of things about it; I hadn't studied it a lot.

And then I read that book by Ivan Sanderson and that kind of opened my eyes a little bit, and then things just kind of parlayed from there. Until Roger asked me to take him down to Northern California to Bluff Creek, as they had found footprints around that area from them new roads they were building down there for logging. In fact, that's why I was there; Roger asked me to take him down there because they found those footprints there in late August.

LCS: In 1967, you and Roger Patterson captured the most famous film evidence of bigfoot to date, known as the Patterson/Gimlin Film. Are you surprised at the level of controversy and notoriety this film has caused?

BG: Oh by far. Yes, I had no idea at the time. Had no idea.

LCS: Are you surprised that no one else has managed to capture film or video that clear since then?

BG: That really surprises me. I just can hardly imagine that all these different sightings that people talk about they've seen, why they weren't able to get film or pictures of them. Yeah, I'm really surprised, that just amazes me.

LCS: Do you think the film is more popular today? Do you think people talk more about it today compared to back in the 1960s?

BG: In my opinion, yes. Of course I didn't know as much as I know now. I didn't think there was as many researchers out there—you know, trying to get evidence; and actually the people have come forth and talked about the sightings and encounters that they've had with these creatures. I never knew that much, not even one tenth of that before. So I'd say, yeah, there's a lot more interested people than ever before.

LCS: Everyone has seen that famous piece of footage, but it is very short. Tell me, what events led up to what we see in that film, and what happened afterwards?

BG: Well, it was a nice fall day, sunshine and everything, and Roger was taking pictures of the scenery and of me riding and the leaves; cause you know all the leaves were turning pretty red colors. We never dreamt we was going to see anything when we came around the bend near this down felled tree in the creek there, and this thing just stood there. And then you see everything in the film footage. After that, Roger got the pack horse caught back up and then he wanted to go and get the plaster of paris, about two to three miles back to the camp and come back.

Of course, by the time he got the plaster casts poured and ready to go, it was practically dark; the sun sets low the last part of October, so by the time we got back to camp, we got in the truck and went into Willow Creek. Roger wanted to get a hold of Hodgson, and call the people in Canada with the track dogs and see if they could track it up into the woods. Then I drove him on over to mail the film. I thought he mailed them to Yakima and on to Al DeAtley, that's what I understood, but I never went in with him so I don't really know where he mailed them to.

Note: After the film was taken the two men tracked the creature until they could not follow it any further, as it went up a mountain. They then returned to the film site. Also, Bob's reference to mailing actually means "shipping" as post offices were closed and we believe the film (or films) was shipped by an air carrier service. There were actually two film rolls and we believe they were both shipped together.

LCS: I keep hearing that Roger rented a camera. I know it was a Kodak camera, 16-millimeter film. Do you remember how many rolls of film you guys had? If I'm not mistaken, didn't the film run out? And he had to put a whole new roll of film in it? Because I know a roll of this film doesn't last that long.

BG: You know, I don't know that much about it as I didn't have nothing to do with cameras or anything at that time. I still don't take pictures. But I do know he had to have one other roll as he had to put another roll in the camera.

LCS: Were you still involved after the film was shot and Roger was showing it to all the different places. Or did Roger take it and become involved in his own thing?

BG: Yeah, Roger and Al DeAtley took it and did their own thing with it. In fact, I went back to work with my job and they wouldn't answer my phone calls or questions about what was going on with anything. So they alienated me completely out of it.

Note: What Bob means here is that he quickly became alienated. He did work with Patterson and DeAtley initially in public presentations of the film.

LCS: Obviously, Roger was selling his book and all. Did he go to bookstores for signings? Giving autographed copies to people? Did he ever at least want you to be present? Did he ever do any of that?

BG: Well, I heard that he did. Now before the film was shot, I helped Roger on a book signing deal on his first book. I was there. But after the film, Roger and Al just kind of tried to keep me out of it as much as they possibly could. Once I came back here and I went back to work. As I couldn't afford to stay off work. So whatever they sold after that they wouldn't even let me know what was going on. They had that office there in Yakima; they had three or four women working in there. I'd go talk with them to try to get information on where Roger was and what was going on, and they would deny me any information of any kind.

LCS: Why was that? Did you and Roger get into an argument? Why did he act that way?

BG: Well, I think it was probably Al DeAtley, Roger's brother-in-law. He was running the show, and he wanted the money out of it; he didn't want to share anything with me. You know, I was supposed to be one-third partner, you see? But I could never catch up with those guys 'cause I was trying to keep down a job.

LCS: How do you feel when you hear Bob Heironimus is still around saying he was the guy in the suit? And that Roger agreed to pay him one thousand dollars to do it? What is your opinion of Bob Heironimus?

BG: Well, ya know, I think it's all money. If they think they can make a few bucks on it, they'll say anything.

LCS: In your opinion, did Roger ever ask Bob Heironimus or anyone to do something like that?

BG: Well, I don't know what Roger did. But if he did, I never knew about it.

LCS: What's the one question you're the most sick and tired of being asked?

BG: Well, I don't know really. Well, "was it real?" The question I get tired of being asked is, "Was it a man in a suit or was it real?" That's mainly what I get tired of being asked.

LCS: Did you and Patterson ever return to Bluff Creek to try and get more footage? Or perhaps bring cameras to different areas?

BG: Well, I don't know about Roger, but I was never there before, and I never was there afterwards, until 2003 for the Willow Creek conference.

LCS: Have you had any other sightings or personal encounters since that famous one in Bluff Creek?

BG: No, I haven't.

LCS: Some people say they believe in bigfoot, but hope they're never discovered. That some things are better left mysteries. How do you feel about that?

BG: Well, I sure think it would settle a lot of questions, although I almost tend to believe the belief of the Native Americans, that they ought to just be left alone.

LCS: Let's just say that someone found a dead bigfoot alongside the road; do you think it should be brought in?

BG: Oh, definitely I do. Yes I do. If that was the case I'd like to see that. But I don't want to see nobody shoot one.

LCS: If you were out in the woods and you found one dead, what is the first thing that you would personally do?

BG: I would walk away from it and get away from it as fast as I could. I went through so much crap with this film footage that I don't want nothing else to do with it. You know I don't want to go through the same old thing again.

LCS: Thank you very much for your time sir. It was an honor.

BG: Have a good one.

(Left to Right) Bob Gimlin, John Green, Thomas Steenburg, and Dmitri Bayanov on the road to the Willow Creek Bigfoot Symposium in 2003.

JOHN D. PICKERING
March 2007

John D. Pickering

LCS: Hello John. Can you tell us about yourself and how you came to be interested in cryptozoology?

JP: I'm forty-one years old and have lived in Grays Harbor County, Washington my entire life. From a very early age, the outdoors played an important part in my life. My father was an avid fisherman and hunter. Much of his off time was spent in the outdoors. From the time I could walk we were fishing and exploring. While in junior high, my father partly retired and we had the summer times off together. We would go camping after school got out and wouldn't go back home until a few weeks before school started. This gave me great time to explore.

It was during a hike that something strange happened that I just could never explain. This made me wonder for years about the possibility of the existence of sasquatch. This wonder grew for years. I decided to look into the subject and was amazed to find out how many people were interested in it and how many people claimed to have had similar experiences.

LCS: Tell us about the NAAP and how you became involved with it.

JP: NAAP is short for the North American Ape Project. It is an ongoing research project looking for evidence of the existence of a North American Great Ape. It is directed by Dr. Jeff Meldrum, anthropology professor at Idaho State University. The good doctor was doing some field work in my area, and one of the people who was helping him knew me and knew that I had good working knowledge of the areas they were interested in. He introduced Dr. Meldrum to me and after some time, I became a volunteer with the project. The project has numerous members in several locations who conduct ongoing field work.

LCS: What is the best evidence in your opinion of the existence of bigfoot? Why are you so convinced?

JP: The numbers of track castings from so many locations over such a wide period of time to me represents a good case for something being out there. To think they are all hoaxed is harder for me to believe than to think that there may be a real animal out there making them. Of course there are hoaxes, and unfortunately that makes it harder to make the case that these animals are real. I also have talked to many people who claim to have seen these animals. Some of them are people who I personally know and have little doubt that they are not mistaken about what they saw.

LCS: Tell us about some of the tools and techniques that you employ during a research expedition. What kind of equipment do you use to try to find a bigfoot?

JP: I try to take a more passive approach to research. I try to create as little of an impact in an area I'm working as possible. I've used remote cameras in some areas, as well as night vision equipment. I look for an area that has a history of tracks and/or sightings. Basically I really like to just get a good feel of what is going on in an area. What animals are using it, what they are eating, when they are moving. Then I try and pick locations to set cameras, or sit and watch and listen. Then record what I see and hear. This isn't exciting and is a long-term approach. Hopefully, over time, I can come up with enough information that may give a small piece of the puzzle that opens up a window to some better discoveries.

LCS: Are you an advocate of techniques such as "call blasting" and "wood knocking?" Some people even believe that singing a soothing melody might lower the guard of bigfoot and make them less fearful of you.

JP: I have mixed feeling about these techniques. I think it really has to do with what you're looking to do. If you are out to get some sort of evidence as quickly as possible, there may be some merit to these techniques. Call blasting is a tool that shows great possibilities. The biggest problem I see is we just don't know for sure what a sasquatch is and what sounds may be real and what may be something else. We also don't know what the sounds mean.

If you broadcast the wrong sound, you may just be alerting the sasquatch that you are there. If you broadcast the right sound and attract a sasquatch, you need to be prepared. You may only get one chance. If the big guy shows up and sees a group of humans, he may never fall for the trick again. While you're fumbling around for the night vision and a camera, your great call-blasting success just hit the next ridge over, and all you have to show for it is a group of people saying that it worked. Well, at least once.

The other one that happens is the call is broadcast and there is a return call just before the recording unit is turned on. I guess what I'm trying to say is, if you are going to do this, you need to think it through and be prepared for anything. More times then not, nothing will happen. But then again, we aren't doing so good at getting good evidence so far.

LCS: Have you ever had a sighting or any kind of personal encounter with a bigfoot/sasquatch?

JP: My friend and I hiked into a remote lake in the southern Olympics to do some fishing for a couple of days. The first day we heard crashing noises coming from the woods, but didn't think too much of it. We figured it was elk or bear. At the camp that night, when it got dark, the noises became more frequent and much closer. The sounds of large tree limbs being broke and crashing noises would start up from a dead silence, go on for several seconds, and then nothing but silence again.

Half an hour or so would go by, and it would happen again, but from another direction. A couple of times, there was the sound of something being thrown through the brush. This kept up all night until it got light. As soon as we could see well enough, we got out of there.

This haunted me for years until I just couldn't stand it anymore, and I had to look into what could have done this that night. Since then, I have done much serious investigating.

During these times, I have been roared at by something so powerful that I know of no known animal that could make such a noise. One roar in particular happened in the middle of the day while hunting deer. My partner and I were heading across a logged-off area back to my truck for lunch. We were about fifty feet apart. A stand of big timber stood 800 yards in front of us.

All of a sudden, out of the timber came what I can only describe as an extremely powerful roar. It lasted for about five seconds. It was so powerful and loud that I knew right away it wasn't made by anything I was aware of. My friend looked at me and we both said at the same time, "What the hell was that?" Then it roared again for about three seconds. A few seconds later, came another three-second roar.

We headed for the truck wondering what made the sounds. We got to the truck and my friend said, "Maybe we don't want to know what did that!" But I was pretty sure I knew what had.

A week after the roars, I was looking around behind that patch of timber and found what appeared to be some sasquatch tracks. They were full of rainwater and, after draining them, it was clear the water had destroyed any detail for casting. A week after that, I was in the same location just as darkness fell and heard some very clear wood knocking sounds just up the ridge from me. I waited for some time, but that was all that happened.

I have had my dog do things that were very out of the norm while I believe I may have been in the same area as a sasquatch. I have heard many stories over the years and I've seen a lot of odd things that really make one wonder.

LCS: If a person comes face to face with a bigfoot out in the wild, what action do you recommend they take? Should you freeze and do nothing? Should you smile and try to be friendly? Should you turn and slowly walk away? What would you do?

JP: Well this is hard to answer. Every situation is a little different. But I would say the best answer would be to stay still. Take a non-threatening posture and keep your eyes down. Do not stare and do not show your teeth. Most animals take this as aggression. If there doesn't appear to be a threat, I would just stay that way and carefully try and observe the sasquatch. Of course if you feel you are in danger, I would say to slowly back away and leave the area.

LCS: What would you speculate is the history of this animal? Are they descendants of *Gigantopithecus?* Are they "The Missing Link?" Are they something else entirely?

JP: Well that truly would be speculation. I think the giganto theory has some merit. The problem is we just don't have enough of the pieces of the puzzle, on the giganto side or the sasquatch side.

LCS: What do you think about the theories some people have that these creatures have some sort of mystical powers, exist in a different dimension, or have something to do with UFOs?

JP: I think people are allowed their own theories and ideas. I don't have the answers. Personally, I haven't seen or read anything that would connect sasquatch with these ideas. I don't focus my research in that direction.

LCS: Why haven't we found them yet? Why hasn't that clear picture been taken? Why haven't we found a body? What are we doing wrong?

JP: Well I would partly say we have found them. Thousands of eyewitnesses and footprints tell us that. Clear pictures are easy to explain to me. Many people don' t carry a camera with them. Those that do may have it with them but it is stored away in a camera case or bag. Now, if you take the majority of witness reports of a sasquatch, it only lasts a few seconds and the animal is gone.

Now assume you are the average hiker or motorist and you are not expecting to see a sasquatch. You see it and your first thought is to identify it as something you know—man, bear, whatever. Then by the time you go through the list of what it could be, and you come to the conclusion that it may be a sasquatch, it is usually gone. Then in a couple of minutes, you remember that you have a camera. Now you feel like the biggest idiot in the world because you just saw a legend and didn't get a picture.

Another thought is that many people have a hard time taking a good, clear picture of a flower in the bright sun. Now add a moving object at distance, low light, and enough adrenaline to cause an earthquake.

Bodies are a lot the same as a photo. First of all, most people aren't looking. A body in the woods gets eaten, scattered, and decomposed fairly quickly. If you see large bones scattered around in the woods and aren't really up on your anatomy, you probably would just think it was an elk or some other large animal. Unless you found a skull you wouldn't think that much about it.

Now given that we are talking about a rare, secretive animal, the chances really get bad. There are several thousand cougar in Washington State. Naturally dead cougar bones are next to impossible to find. I know as many or more people who claim to have seen a sasquatch than a cougar.

I don't think we are really doing anything wrong, other than not spending the money to put a scientific research project into effect that would put a good group of people into the field fulltime to find the evidence. Weekend warriors are just not enough.

LCS: What do you think will happen if and when a bigfoot is captured, or a dead body found? Do you think the animals will be doomed, or will we somehow be able to protect them?

JP: I really don't know what would happen, but I would think there will be some knee-jerk reaction for a while. First it will take the scientific community some time to sort out what it is. Even if a body is right in front of many in the scientific community, they will claim it is some sort of hoax and refuse to look at it. Others will claim it is some sort of freak and that sasquatch doesn't exist. After a while, the majority will accept that they must be real and call for some sort of study. This is when the legal nightmare will start.

You will have hard-core environmental groups calling for the complete closure of all forestlands to protect the endangered animal, while the timber industry says they don't exist. This will be battled out in the courts for years. Personally, I don't think either side will be right. I don't think they are endangered or that logging hurts them. In fact, I would guess that like many larger animals in the Pacific Northwest, the sasquatch benefits from good forestry. Eventually it will get worked out and the sasquatch will be in pretty much the same boat as they always have been.

LCS: The majority of bigfoot researchers are not paid, and investigate on their own time. Everyone agrees the only thing stopping us from getting that irrefutable evidence is funding, and lots of it. So why is it that bigfoot research organizations haven't reached out to the billionaires of the world to try and get this needed money?

JP: I suspect some have. In fact I know that some have. The problem is that the general public doesn't really know much about sasquatch research. The media usually plays it down to make it look like a bunch of nuts are the only people who take it serious. Add the known hoaxing and the skeptical eye the mainstream scientific community has, and you make sasquatch research look like a losing proposition.

LCS: Any final thoughts? Anything to say to the nay-sayers?

JP: I'm the last person to say that I know what a sasquatch is. Or that I can prove they exist. We don't have all the answers. But when faced with hundreds of years of stories, sightings, tracks, hair and screams in the night, it is hard to just say that there isn't something to it. Is it good science to turn your back on hundreds of years of data and information, to refuse to look at it? Or is it good science to work to get to the bottom of it and learn the truth?

LCS: Thanks very much, sir. Good luck finding our big, hairy friends!

JP: Thank you!

RAY CROWE
April 2007

Ray Crowe

LCS: Thank you for taking the time to speak with us, Mr. Crowe. Can you tell me about your background, and how you came to be interested in the research of bigfoot/sasquatch?

RC: Born December 30, 1937, Portland, Oregon. Went to school at Northwest Business College and Portland State University, and served in the Air Force in weather observation. Worked technical jobs at paper mills and Tektronix, had an antique shop and a used bookshop. Book shop customers, Native Americans, got me interested in bigfoot; convinced me to start a society in Portland, which I did in 1991. We had outings, meetings, a website, and a newsletter *The Track Record;* now working on issue 170. Have not seen a bigfoot, but found tracks numerous times, suspected hair, suspected vocalizations, broken trees, and strangely positioned rocks.

LCS: You have a unique and somewhat controversial theory regarding the origins of these animals and how they came to live in North America. Tell us about it.

RC: I'm one of the few that believe bigfoot is descended from *Homo erectus*/Neanderthal stock. Opposition from many says there is no evidence of culture in bigfoot. I counter with loss of culture by Tasmanian natives; early modern man hunted and killed off most bigfeet (*Homo erectus*/Neanderthal), driving them to near extinction; and saying that early man killed off *Gigantopithecus* until extinct. Height is the result of general late Pleistocene gigantism of many species. I believe that early man was, and still is, hairy. DNA evidence is not all in, but so far, indicates bigfoot is of human stock.

LCS: You have quite an extensive collection of bigfoot artifacts and memorabilia. What is your favorite item in your possession, or the one you find most intriguing?

RC: I have a stone effigy shaped like an ape found in an excavation for a house in Oregon many years ago. Also, have an early Mexican clay dinosaur given to my missionary parents; plan to donate to an upcoming museum exhibit soon, along with other artifacts.

LCS: Who in the field of sasquatch research do you admire the most? Whose contributions do you find to be the most groundbreaking, and has the best chance of making "the big discovery?

RC: I believe the efforts of Tom Biscardi will yield the first specimen. His crew is very active and has the most modern equipment for the task. He is not a "sit at home" investigator, but spends a great deal of time in the field tracking down leads. Yes, he has made some errors so far, but then again, so have I.

LCS: What is the oldest bigfoot story that you know of in the Pacific Northwest that you believe to be true?

RC: Probably the Ostman story; that is, the prospector in the 1930s who was kidnapped by a sasquatch in British Columbia. The story had a wealth of detail ascribed to the creatures that I find fascinating. Many stories appear "good" to the point of being hard to decide on—Roosevelt's Bauman tale, for instance.

LCS: What is the most frightening bigfoot story that you know of?

RC: It was reported that a bigfoot attacked a woman in the Olympic National Park, tearing her limb from limb. The tale was told by a dog rescue handler, and came from a popular bar, and might not be true. There are enough other incidents of bigfoot killing people to add weight to the story: the killing of a hiker near Sisters, Oregon; the killing of a family at picnic spot near the Canada border in Washington; bent rifles found along with missing hunters in two instances; the killing of miners along the Chetco River, Oregon; and the Bauman story.

LCS: Do you think there are different sub-species of bigfoot, depending on where they are? Are the bigfoot in Washington State different from those in Texas, Ohio, or Oklahoma?

RC: I have suspected, due to different descriptions, that there are varieties. I suspect that *Homo erectus* migrated along Pacific waters from Asia into North America, following game as they migrated across Beringia. They could have swum (possibly still do) the straits, come in crude boats along the shoreline, or walked on dry land. There is some evidence that European Neanderthal man could have crossed the Atlantic, following seals, etc. Again, either by swimming or crude boat, arriving along the Atlantic coast of North America.

At some time, the creatures could have met and interbred. Also, there are many tales of their breeding with Native women, and leaving living, viable hybrids. All of the former could account for the reports of tall and short, flat and pointed headed, docile and violent, creatures.

LCS: Most experts agree that there are many species on this planet that have yet to be discovered and documented by science. Gorillas were not even "discovered" to be real until 1902. Do you think that besides the sasquatch, there are possibly other primate species yet to be discovered?

RC: Definitely. As with the early men surviving in small pockets, the same could easily be true of several Asian, African, and South American reports. The niches for their existence is slowly shrinking, and even today there are new large mammals frequently being described by the sciences. Then, there are many accounts of the "little people" that need to be investigated also.

LCS: What is your opinion of other cryptids usually mentioned in the same breath as "bigfoot"? Such as the Loch Ness monster, Mothman, and chupacabra?

RC: I'm not really familiar with these other creatures. I often wonder about aquatic reports being large sturgeon or eels, and others as erroneous descriptions of bigfoot, like the Mothman. Other than that, I keep an open mind, and just let the reports stack up. I'm quite sure, especially in the oceans, that there are many, as yet, undescribed species to be found.

LCS: What are your thoughts on the migration habits of sasquatch? Do they migrate? Is it possible they might travel because of weather, the winter cold, food sources, the pursuit of mates?

RC: If food is reliable, the family unit will stay in place year round (lots of tracks found in snow in winter), but...seems that singles, both male and female, go out searching for mates. One record by Mike Jay in Coos County, tells of contacting a fellow by the name of Haas in California with a query about a track that was peculiar, a split, and he wrote back with a photo "Is this the track?" It was from Northern California, miles away from Coos County. If food is scarce, some of the individuals might well migrate south in the winter, possibly following or swimming in main rivers along the way—especially in the midwest and east. My opinion.

LCS: Some scientists and authors theorize that bigfoot actually has some sort of psychic, or extraterrestrial connection. That they are in some way mystical or psychic beings, or that the bigfoot phenomenon is directly correlated to the phenomenon of UFOs. What are your thoughts on these theories?

RC: I have listened to and taken many, many strange reports where tracks disappear, or the creature disappears into a ball of light, or was seen in contact with a UFO. For the time being, I prefer to stay open-minded and sit on the fence. I have a report where a bigfoot leapt thirty feet. Disappearing tracks? Reports of their being exceptionally quick, disappearing when one glances away? I tend to keep my "skepticals" on with all reports, but the UFO ones are the strangest. Yes, I believe in life on other worlds, so, I guess it's possible. But for the time being, I'll continue to collect data, be skeptical, and again, let the reports stack up.

LCS: What do you think is the biggest mistake made by researchers today?

RC: In-fighting with other researchers, hogging information, not spending time in the field, getting lost in "tunnel vision," not reading the many reports with an open mind.

LCS: If you have one piece of advice for future generations of bigfoot researchers, what would it be?

RC: Stay in the field as much as possible, especially in the same area, let the creatures get used to you, do baiting, look for anything unusual—like strangely placed sticks or stones—keep notes, don't make up your mind on anything a hundred percent. Leave lots of room for new ideas and theories. People have been hit by rocks, for instance.

LCS: Any final statement you'd like to make? Anything to say to the skeptics out there who might call you nasty names and denigrate the work that you do?

RC: I have learned to live with it. Sooner or later, one of these creatures will be captured, and many of the questions will be answered. Until then, it's better to be in the group that said, "maybe you're right" rather than to be proven wrong. Visit my website [www.internationalbigfootsociety.com]; it has over three thousand reports to date, even the radical ones, not just those that support your particular theories. Even John Green collected UFO and other strange reports.

LCS: Thank you very much for your time, sir. It's been a pleasure. Good luck in the search!

SEAN FRIES
April 2007

Sean Fries

LCS: First of all, thank you for taking the time to share your thoughts with us, Sean. Can you tell our readers a bit about yourself and how you came to be interested in the bigfoot/sasquatch phenomenon?

SF: I am forty years old and I live and work in Trinity County, California. I have been involved in bigfoot research for about eight years or so, formally. I am the lead field investigator for the Bigfoot Research Project, and that was the group that was run by Peter Byrne. He has retired from the field and is part of a wildlife conservation center in Tibet. He still lives in Southern California.

I became interested in the sasquatch legend when I was very young. One of my dad's friends was the son-in-law to Archie Buckley, who was one of the "Bay area four," and so I was always being told about the big hairy man that lives in the woods. Then in 1988 while I was home on leave from the military, I went backpacking down into the middle fork of the Feather River and had an encounter that is pretty well documented. I was stalked by more than one of these things and was spooked pretty badly. Well, I finally decided to tell someone what happened and that's how I got involved with research.

LCS: Can you tell us about the Bigfoot Research Project?

SF: The Bigfoot Research Project was started by a man named Peter Byrne. He was originally associated with the Tom Slick expeditions that tried to capture one of these creatures and also the hunt for the Yeti in the Himalayas. He has been on all the old *In Search Of* TV shows. Anyway, he retired to start a nature preserve in Tibet and the Bigfoot Research Project went dormant for several years. I was part of a research group called the Bigfoot Ranger Team and that was run by a man named Ian Garrott.

That group was having some internal problems and another of my friends had just joined the BFRO, so essentially I was left on my own. I decided to look up Peter Byrne and see if he was still active, and maybe if I could work for him. After I contacted him, I found out about him not being involved in the study of sasquatch any longer, and the whole wildlife preserve in Tibet. I asked him if he would mind if I picked up the banner and ran with it, and so I took over the field investigations of the BRP.

LCS: What do you think is the most common misconception about bigfoot?

SF: I guess it's that people don't understand why we haven't killed one, found a body, or even gotten another Patterson film yet. I also hear about the whole Bob Heironimus angle all the time where people say, "Well, didn't that guy come out and say he was the man in the suit?"

LCS: Many techniques are used by researchers to draw in a bigfoot, from call blasting, to setting up food as bait, to placing hidden cameras and more. What techniques do you find most effective?

SF: Call blasting hardly ever works. I have seen it work when done in a scientific way, but for most people it just does not draw in a creature. I like to think of it like this: We are looking for a moving needle in a haystack (although I have found a pattern of activity where I know what areas these creatures are most likely going to be in during a certain time). Camera traps have failed to capture a single image so far. (Richard Noll caught something on his own up in Washington State, but it was not a clear picture.)

These are highly intelligent beings, and they are very shy and wary of anything new in their environment. We have had camera traps moved and in some cases smashed to bits, so they are a waste of time and energy. I feel that in order to have an encounter with these creatures, you have to be out there with them and compete or be seen as competition for the resources such as food, water, etc. Then they will come out and challenge you or try to scare you off by rock throwing, screaming, and these other types of intimidation.

LCS: Have you ever had a sighting or personal encounter with a bigfoot? If so, what happened?

SF: My first encounter is well documented. It took place in June of 1988, North Fork of California's Feather River in Plumas County.

After making camp and cooking a few trout I had caught earlier, I was getting a little tired, so I decided to turn in. I climbed into my tent and lay down on my bedroll. I let my dogs run around because they always stay close to camp. I started to dose off to the crickets chirping when suddenly I woke up. It was as if I had one of those dreams where you are falling. I could tell there was something very wrong. It was dead quiet—no crickets, nothing; and my dogs came running into my tent shaking. These dogs were usually very aggressive.

I grabbed my rifle and pistol along with a flashlight, and stepped outside the tent. I couldn't see anything, but I had that sensation of being watched. I grabbed some more firewood and threw it on the embers left from the dinner fire. Then I heard some very heavy footsteps right behind me in the trees. There was also a very strange odor, almost like a cross between a skunk and something dead.

I stayed up all night long by the fire. There was more than one of them. I could hear one walking around the tree line of my camp and I could follow where it was by the sound. It would stop, I would focus on that spot, then there would be another one walking in another area of the tree line. These things circled my campsite all night long.

At first light, I broke camp and got the hell out of there. I was still three days from being picked up over in the town of Quincy. The whole next day I felt like we were being followed and my dogs never left my side. I was pretty freaked out and on edge. A few times during the hike, I could swear I saw something moving in the trees about 100 yards away. I am now a scientist/bigfoot researcher. This episode on the Feather River was life changing for me, and I found myself on a forever quest for the elusive bigfoot.

LCS: Are you of the "kill" or "no kill" mentality? Can we capture one without killing it, and should we even try to capture one?

SF: I think that the only way that mainstream science is ever going to accept these creatures, is someone is going to have to throw a body at the feet of science. I personally would not pull the trigger because to me, these things are not just some dumb animals. As far as capture, I would hate to see one of these creatures in a zoo.

LCS: Can you make any speculations about the societal structure of a family of bigfoot? For instance, their practices of child rearing, or customs?

SF: Okay, here's my theory and it's controversial. I have spoken with a lot of Native Americans about these things, and they all without exception consider these things as another tribe of Indians called "Seeahtik." I think that they just might be another type of hominid. Not human like us, but something really close; so close that our DNA looks just like theirs.

LCS: I hear people say all the time, "If they exist, why haven't we found them yet? With all the satellite tracking, infra-red, night vision, etc., surely with this technology we should have obtained proof." Well, with all the technology we have, why haven't we obtained that proof yet?

SF: Well, if we had the resources to have all that technology at our use, it would be over with really quickly. It costs a lot of money for thermal imaging and a dedicated satellite. That would be way too expensive. If I was to win the lottery today and had a large sum of money at my disposal, I would venture to say that this mystery would be over in short order.

LCS: What have we been doing wrong? Do you have any theories about new techniques we could try that might get us better results?

SF: Well as I said before, it comes down to being able to compete with them for the natural resources. These BFRO outings where they take fifty or a hundred people out in the woods are not going to work. You put fifty people in an area of the woods and you would be lucky to see a squirrel, let alone a bigfoot.

LCS: Sometimes a bigfoot will quietly sneak into someone's campground. Sometimes they "shadow" people as they walk, taunting them. Other times, they are very threatening, throwing rocks and sticks, grunting and screaming. Do you think this is just the personality of the bigfoot, the importance of the territory, or reaction to something the humans are doing?

SF: I think they are curious to a point, but it comes back to the whole competing for resources idea. If we are in an area where they want us out they will try the intimidation tactics to make us leave.

LCS: What is the most burning question in your mind that remains unanswered? What do you want to learn about them above all else?

SF: Just what the hell these things are. Primate, hominid, giganto or whatever.

LCS: What do you say to the skeptics when they say, "That bigfoot stuff is all a hoax. The guy who did it even admitted so on his death bed."

SF: People believe without question everything they see on TV or hear on the radio. People do not know that Ray Wallace changed his story like he changed his underwear. So to me, that makes any of the things he said not credible. I do not worry about what people think when it comes to my work in this field; it's all a personal journey that I am on, and I am not doing this to get rich or famous. And I have yet to be called crazy or stupid. Most people are very interested in it.

LCS: Thanks a lot for your time, Sean. We very much appreciate it. Keep us posted.

SF: It was a pleasure, and by all means, feel free to look me up sometime if you want to go out in the field.

MICHAEL RUGG
July 2007

Michael Rugg

LCS: Hello Mr. Rugg. Can you tell our readers a bit about yourself and what you do?

MR: I am the curator of the Bigfoot Discovery Museum in Felton, California.

LCS: Tell me about your bigfoot museum. Where exactly is it?

MR: It is located in the Santa Cruz Mountains, near Henry Cowell Redwoods State Park. It includes exhibits on popular culture, local bigfoot sightings, the Patterson/Gimlin Film and bipedal primates in the fossil record. There are casts of footprints, handprints, and a heel print from the Skookum Cast. Our "main attraction" is a nocturnal diorama featuring bigfoot models that once resided in a museum in Seattle back in the 1970s.

LCS: When did this fascination with cryptozoology begin for you?

MR: It started back in 1950 when I had a sighting in Humboldt County, Northern California. I began collecting in 1951 when the footprints of the "abominable snowman" were photographed on Everest.

LCS: You had a personal encounter? Can you tell us about it?

MR: Yes, I saw one as a child in 1950, in Humboldt County, California, on the South Fork of the Eel River. My dad once had a sawmill in Laytonville and my mom was an avid driftwood collector. That's why we went to Humboldt on camp-outs from time to time. On one occasion, my parents were preoccupied with the preparation of breakfast, and didn't notice me follow the trail out of camp toward the river.

I stepped through some tall reeds out onto a sandbar, and there stood a very large hairy man, wearing nothing but an ill-fitting torn shirt. As I stared at this big, hairy man, my parents, realizing their five-year-old had wandered off, started screaming my name. Not wanting to take my eyes off the "wildman," I finally turned and ran back to the camp. When I convinced my parents to follow me back to the sandbar where I had seen the creature, we discovered it had gone. My parents wrote it off as a "tramp" and advised me to "forget about it—don't worry about it."

LCS: What is the most compelling evidence in your opinion that convinces you these animals are real?

MR: My own sighting and the hundreds of stories I've heard from people who have visited my museum and shared their bigfoot sightings and encounters, and the Patterson/Gimlin Film.

LCS: You are also an artist? Tell us about some of the bigfoot-related artwork you do.

MR: I started the Bigfoot Discovery Project back in 2003, and we opened our doors at the Discovery Museum in 2004. So many people came in with local sightings, that my art projects have taken a back seat to local fieldwork, our newsletter, and development of the museum premises. My first completed project was a coloring and activity book aimed at younger seekers.

There are posters in the works, and I'm gearing up to produce some sculptural interpretations as well. I'm also interested in assembling a tour book on our local bigfoot, and a DVD about our museum. There are audio-video projects on the drawing board as well, which will be available for perusal on interactive kiosks at the museum in the future.

LCS: In your years of research, what are some of the most common misconceptions people have about these animals?

MR: First, that there is only one sasquatch, named "Bigfoot." This would of course place it directly into the same realm as Santa Claus. Second, that there was a "deathbed confession" that the Patterson/Gimlin film was a hoax. That never happened. Third, that they are strict vegetarians— that misconception probably arose as a result of the film *Harry and the Hendersons.*

LCS: Investigators have reportedly found samples of hair and scat that yielded samples of DNA, which did not match humans or any other known animals. Why is this still not accepted as proof?

MR: Because "science" requires a "type specimen" with which to compare things; "science" also requires repeatability, so once a sample animal *[DNA]* is "collected," they'll (the "scientists") still want to see another before they'll all be happy.

LCS: Scientists and experts have learned (and agreed on) much about these creatures through deduction and educated speculation, answering many questions. What are some of the important questions that still remain?

MR: How about, "What the heck are they?"

LCS: What do you say when you get into the inevitable argument with a cynic or skeptic?

MR: I remind them that people all over the globe have reported these animals since the beginning of recorded history. I assure them that the Patterson/Gimlin film has never been debunked. And I point to all the examples of ape-like bipedal hominids in the fossil record. I also point out the arrogance of someone who has never seen one of these animals, to scorn all the thousands of eyewitnesses, and with a wave of the hand, relegate them all as fools or liars.

LCS: Thanks very much for your time, sir. The best of luck to you.

MR: Keep on the right track.

Image on the following page

Among his many talents, Mike Rugg is a remarkable artist. He created this scene called *The Moment* which shows the moment Roger Patterson and Bob Gimlin spotted the creature they filmed at Bluff Creek on October 20, 1967. Mike consulted with Bob Gimlin, Chris Murphy, and other researchers to ensure details were correct. Even the men's clothing depicts what they were wearing that day.

RAY ROSA

July 2007

Ray Rosa

LCS: Hello Ray, thanks for joining us. Can you tell our readers a bit about yourself and how you first became interested in bigfoot?

RR: I am a US Navy veteran, and I have been researching since 1988 when I had my first sighting in Bald Hills, Washington, of a large sasquatch running down the mountain yelling and "whooping" as it ran by me.

LCS: What was it that first convinced you that these creatures are real?

RR: Number one, the Roger Patterson film, plus the fact that I have had several sightings in my nineteen years of research, and the overwhelming evidence we have collected, as well as other research groups and numerous sightings.

LCS: What is the best piece of evidence you've collected in your years of research?

RR: My collection of foot casts, hand casts, photos of tracks, and a partial bigfoot photo.

LCS: What kind of personal bigfoot encounters have you had?

RR: I have heard several vocalizations, screams, and three sightings, including an albino bigfoot.

LCS: Countless people have tried and failed to obtain "definitive" evidence of this creature's existence. Can you think of any new techniques that no one's tried yet to collect evidence that might yield better results?

RR: I suggest finding an area where bigfoot frequents. Perform your research in that area so they are used to you and provide a baiting area with food, and if you gain their trust enough, you might see them.

LCS: Some people say they believe in bigfoot, but hope they're never discovered. That some things are better left mysteries. How do you feel about that?

RR: I agree to the point of us encroaching further into the forests. Proving sasquatch/bigfoot's existence will ultimately help save their environment and protect the species.

LCS: What do you think it would do to the scientific community and the world in general if bigfoot were publicly acknowledged as being a real species?

RR: For the skeptics, it would be an embarrassment to know they were wrong, and for the believers a continuation to the study of bigfoot with real-world financial backing as compared to private funding.

LCS: What is your opinion of other crypto-animals or "unexplained" phenomena, such as the Loch Ness monster, chupacabra, Mothman, aliens, or ghosts?

RR: It is my feeling other crypto-animals and unexplained phenomenons should be researched just as well as bigfoot. There are several groups currently studying these other phenomena.

LCS: What is the one question you have about bigfoot that you most want to have answered? What question burns in your mind more than any other?

RR: How can they be so elusive for so long and be seen by so many, and still remain hidden from our society, especially in modern times?

LCS: When the day comes that they are discovered, what will be next for Ray Rosa?

RR: A continued study into the lives of bigfoot, including day-to-day rituals, habitat, diet, etc.

LCS: Do you have any final thoughts in closing? Anything to say to the cynics and skeptics out there?

RR: I welcome skeptics, as several of our researchers were skeptics, including the late Dr. William York, wildlife biologist, who after one time out became a bigfoot team researcher.

LCS: Thanks very much for your time, Ray. It was a pleasure.

RR: Thank you, Happy bigfooting.

ROBERT W. MORGAN
May 2008

Robert W. Morgan

LCS: Hello Robert. You are the president of the AARP (American Anthropological Research Foundation). Can you tell us more about this foundation?

History

RWM: AARF was founded in 1975 as a Florida not-for-profit 501(c) 3 corporation by myself and attorney W. Ted Ernst, Esq., to facilitate, conduct, and augment progressive research into both the new field of crypto-anthropology and those social and physical sciences which complement anthropology to the benefit of humankind. AARF has consistently maintained "good standing" and has never been the subject of redress, review, or audit.

Initially, AARF concentrated on the interests expressed by me in my film proposal *In Search of the World's Mysteries* (which became the TV series *In Search Of*) as the subject of the feature adventure documentary *The Search for Bigfoot,* my inclusion in the record-setting Smithsonian series *Monsters: Myth or Mystery,* and having been featured in *National Wildlife m*agazine which co-sponsored two of his American yeti expeditions, I was invited for guest appearances on a number of television shows (Larry King, Tom Snyder, Howard Cossell, Art Bell, Montel Williams, etc.). I was also invited to lecture at universities and colleges (UCLA, U of Miami, Florida Atlantic, Kent State, Barry and Marietta College, etc.), and at intermediate schools, and many civic associations.

As its research expanded into complementary fields, AARF created a seventeen-member Science Advisory Board that has included such luminaries as: Doctors Grover S. Krantz, WSU; George Agogino, Director of the Paleo-Indian Institute; Carleton S. Coon, Penn State; Boris Sapunov of Mosco;, and S. Dillon Ripley, then Secretary to the Smithsonian. In result, AARF has expressed itself among new and emerging sciences.

Purpose

AARF endorses the concept that that every human being, regardless of race, religion, education, or ethnic origin, has both the absolute right and a duty to seek out for themselves who we are as a species, where did we truly come from, why are we on this earth, and how we can work together in mutual respect and kindness to make this earth all it can be. To help solve these mysteries without presuppositions of any sort, AARF intends to seek out those rare and "hidden" sciences that attempt to separate fact from fiction without prejudice, profit, or predisposition, so they can educate a supportive membership that pursues edification, enlightenment, and opportunity.

LCS: In 1956, while in the United States Navy, you had a face-to-face encounter with a forest giant (as you have named them). Can you tell us about this experience?

RWM: In 1956 I had returned from a Far East cruise while serving aboard the US Navy aircraft carrier the USS *Princeton.* We had come to Bremerton, Washington, where our ship was put into dry dock for repairs. Seeking time alone, I went into a wilderness area in Mason County, Washington. While sitting on a hillside, I heard something moving in the brush up the slope behind me. Not knowing if it was human or a bear, I shouted. I saw patches of dark hair of something running up the slope. At a break in the brush, it halted and turned to look at me. Being a kid from an Ohio steel town, I had not heard of the bigfoot, yet I was staring directly into the eyes of one. He did not threaten me at all; he was as startled as I was. His eyes were intelligent. I don't know who left the area first, but I would bet that it was me.

I tried to report this encounter—I could only describe it then as a gorilla because I had nothing else to compare it with. No one took it seriously.

A few years later, and after some reading, I realized what it had been.

LCS: In your opinion, what is the best piece of evidence for the existence of these animals?

RWM: Speaking only for myself, it was what I have experienced on my own. I have seen them, smelled them, heard them, and encountered them. I have on occasion exchanged verbal and other sound communications that had meaning, and I have achieved a small measure of trust and trade. The longest string of tracks that I am aware of that were authenticated by a qualified scientist (Grover Krantz) were the 161 giant footprints Eliza Moorman and I located that traversed varying terrain and showed active toes movements.

LCS: What are your thoughts on some of their vocalizations and their meanings? Some of it sounding like so much gibberish, do you think they communicate with each other in a language of their own?

RWM: Mandarin Chinese sounds like gibberish to the untrained ear, as English sounds to the Australian Aborigines. I have no doubt whatsoever that the forest giant people have a hyoid bone similar to our own and can form both vowels and consonants. In the first recordings made by Al Berry and Ron Morehead, one can clearly hear the words "You're not welcome." These are especially important because (a) they conveyed a meaning that was appropriate to the situation, and (b) those recording these vocalizations had them in their possession for decades without realizing it.

LCS: What can you speculate about the level of their intelli-

gence? Do they have culture, a social structure? Do they have a sense of family?

RWM: I am not so arrogant to accept that we are the sole primates on this earth who cherish exclusivity. We invented it, and must fight hard to maintain that image. The forest giants obviously have the "special traits" of a humanoid, such as the opposable thumb, the hallux (big toe that permits balanced walking), large buttocks, an articulated speech pattern, and habitually walk erect.

Addressing their culture is difficult; I would opine that because they have no need of fire, no need to cook certain foods to facilitate digestion, no need to use the skins of animals, feathers of birds, and parts of plants woven into clothing, and no need for artificial shelters from the natural elements, that they would be best described as the sole humanoid on earth that live in harmony with their surroundings. We, on the other hand, must drastically modify our surroundings to the extent that it could be said that we live in spite of nature and not with it. And therein lies the conundrum: once humans accept their reality, we are no longer kings and queens of the world.

LCS: What would you speculate is the history of this animal? Are they descendants of *Gigantopithecus blacki*? A hominid? A primate? Or something else entirely?

RWM: I would speculate that we all have a genetic link to a common primate that split at one time. I would suggest that they are certainly of the genus *Homo*.

LCS: Forest giants have been known to hide and watch campers in the forests; they have been known to leave gifts in exchange for food; they have been rumored to watch over and play with children, showing a kind and gentle side. At other times, they "shadow" people as they walk, taunting them. They can be very threatening, throwing rocks and sticks, grunting and screaming. Do you think this is just the personality of the bigfoot, are they protecting territory, or is this reaction to something the humans are doing?

RWM: I would suggest it an error to assume that all forest giants have the same personality traits one to the other. Their reactions to human intrusions and encounters would be dictated by their personal experiences and interactions with humans; and since humans are also complex in how they will react to encounters, it becomes largely unpredictable. However, it has been my experience that they have a high level of curiosity, not so much as something that stems from envy, but is probably the result of previous encounters. If a human acts aggressive, they usually leave the area. If a human acts indifferent, they will either ignore them or, at times, become curious because that may not be their previous experiences. I have found them to be very slow to accept any interaction. Those that do are usually young males. The older ones may have had so many negative encounters that they could care less.

LCS: What do you think about the theories some people have that these creatures have some sort of mystical powers, exist on a separate plane of existence, or have something to do with UFOs?

RWM: Mystical powers are in the eyes of the beholder. And I don't refer to them as "creatures." As to existing on a separate plane, I only need ask in return if people swear they are certain that we all have "souls" that exist after the death of our bodies, why do they deny the existence of "ghosts?" Humans are so damned contradictory.

UFOs are another subject. I once held a responsible position with the FAA in Washington, DC. I know damned well that UFOs have been tracked on our radar systems. Again, as in Operation Bluebook, the military followed the dictates of the elected officials who would not dare use tax dollars to reveal something that would upset the basis of most faith-based religions.

LCS: I hear people say all the time, "If they exist, why haven't we found them yet?" With all the satellite tracking, infra-red, night vision, and technology we now have, why haven't we obtained the proof yet?

RWM: Think about it. How would the innumerable "religious" taxpayers react if their elected officials permitted the use of their tax money to prove that another humanoid exists that would destroy their "faith" that proclaims them to be so very special? They claim that we are "made in the image of God" and that God created everything. It could then be argued by the cynical that perhaps God too did not need clothing to survive the elements, ergo, because the forest giants obviously live in a greater harmony that perhaps...

LCS: What can you tell us about "Bugs the hunter," who he is, and why you would want to speak with him?

RWM: We have an open letter to Bugs on our web site (www.trueseekers.org) which reads as follows (posted in the year 2000):

AN OPEN LETTER TO "BUGS"

> A few years ago I appeared with you on the Art Bell Show. As a somehow frequent guest of Art's show, he asked that I listen to your report of your encounter with the "bigfoot" that had turned both violent and deadly. You and some friends had purportedly killed both a male and a female bigfoot. I was asked to render an opinion both as a recognized authority on the subject and also as a fellow veteran familiar with the stress that can follow combat.
>
> According to your story, you and some buddies had been out running a trap line or such and had come upon two of those entities whom I have come to call the forest giant people. One of your friends had opened fire for reasons of his own. As a Vietnam vet, you had learned the hard way what bullets can do to flesh. Having wounded one of the bigfoot and having sympathy for its suffering, you followed the trail into a dense thicket. There you discovered a female tending to her mortally wounded male companion. When she attempted to defend him, you and your friend opened fire, killing them both.
>
> In examining their bodies, you discovered the facts that I have been maintaining for thirty-five years— they are not apes, they are not mere animals, but they are simply giant hairy folks who are close to us. Out of fear (and perhaps a bit of anger and shame) those bodies were quickly buried at some secret location.

However, in talking with you over Art's show, you appeared to be suffering post-battle stress and remorse. Trust me, my friend, this I understand with all my heart. You appeared both sorry for what you and your friends had done, but it had come to haunt you. Again, I understand. No true warrior feels proud of killing an unarmed person no matter the circumstance.

That night on Art's show you queried me as to why I might think these giant hominids might be more closely related to the genus *Homo* than that of the Pongidae, the great apes. I allowed too that if they were proven to be as I thought, just perhaps they could be included into the family of Hominidae—humans. Under our legal system this could perhaps—perhaps—lead to a charge of second-degree murder or manslaughter.

However, through Art Bell you agreed to give me a map that would lead me to where these two bodies might be buried, remember?

It is my understanding that you had sent half of the map that could lead me to that burial site. I understand too that your wife was fearful for you should I be proven correct.

Several years have passed. I am continually asked by your fans what really happened. Finally, last month I asked Art if he would at least share with me what you sent plus put me into contact with you. He refused—good for him. He is honoring his pledge.

Bugs, IF you are for real, please, please think of the contribution to the world of knowledge that you could make. Please assist me. Please send me the complete map. I promise you that (a) I will keep the actual location a secret and will only permit a limited number of highly qualified scientists to join me on their excavation, and (b) they will be required to sign a non-disclosure agreement as to its location, and (c) I now put you into direct contact with a qualified attorney, Mr. Robert Purser, who will advise and defend you free of charge. You may contact Robert directly at robertp@purserlaw.com.

Bugs, I have to admit that I have received a hell of a lot of "heat" because I bought into your story. Sure, I am human and I make mistakes. Yes, I have been warned by his critics that Art Bell may use such charades to boost his ratings. I am a particularly attractive target because I seldom, if ever, chase sighting reports, and my credibility has a high rating. However, despite his critics, I have continued to trust Art and I have maintained you on the "probability" list on my private chart for measuring reports.

All this said, please know that what you did was: (a) understandable under the conditions, and (b) you would be committing a huge contribution to the knowledge to humankind if you cooperated with me. Please help me— if you are real.

I have asked Robert to NOT contact me about any of your conversations unless you so agree. He will treat you with every confidentiality accorded to any client. Robert, like myself, only wishes to know the truth.

Therefore, Bugs, if you or your wife will please help us solve this very important mystery, please contact us.

Lastly, I would love to have you as our guest on Rob McConnell's "Morgan's Corner" so we can revive what I hope was a trusting relationship.

—Robert W. Morgan
President, AARF

LCS: Do you have any advice for the average guy who wants to go out in the bush to explore and look for signs of a forest giant?

RWM: I explain all that in my *Bigfoot Pocket Field Manual,* which will be available August 1, 2008.
(Note: The book cover and associated information is shown below.)

LCS: Do you have anything to say to the skeptics of the world? Any final thoughts?

RWM: I am not ready to die yet, so my thoughts are not final. And I have nothing to say to skeptics. Their opinions do not alter fact and I could not care less what they think or don't think. They make no difference. For instance, if they suddenly decided that sunrises would not follow sunsets—guess what?

LCS: Thank you my friend. It has been a privilege and an honor. Best wishes to you in all your endeavors!

RWM: Same to you.

In this remarkable book, Robert Morgan explores the mystery of the giant forest people, and this leads him to the mystery of human existence. He takes us far beyond the realm of conventional and orthodox thinking, and reveals some of the secrets of Native American philosophy. Robert was, and continues to be, one of the greatest researchers to take up the sasquatch/bigfoot field of study.

In this work, Robert gives us the benefit of his many years of experience in seeking the giant forest people. It is a very straightforward and sensible guide for those who currently undertake field research, or those who wish to become directly involved in this fascinating study.

Both books shown are published by, and available from, Pine Winds Press, an imprint of Idyll Arbor Inc. http://pinewindspress.com/

TOM BISCARDI
May 2008

Tom Biscardi

LCS: Hello Tom, for the few readers who are not familiar with you, can you please tell us about yourself and how you first became interested in the bigfoot phenomenon?

TB: Sure Linda, back in 1967 I saw the Patterson film on the Johnny Carson show. It changed my life! I was young and sure of myself and thought surely I can go out and find this thing easily. I went to the library (no computers back then) and learned all I could about bigfoot. I wanted to find out what others had done and see if I could identify some of the mistakes they may have made and do it all the right way. Over three decades later, I am still learning, correcting mistakes, and working to solve this mystery. I have had six eye-to-eye encounters and I know that me and my team are right on the cusp of one of the greatest discoveries of our century! I am more excited and passionate about proving to the world about the reality of this creature than I was over thirty years ago!

LCS: You have had encounters with these creatures. What can you tell us about your amazing experiences, and what have you learned?

TB: On each of my encounters, it clarified in my mind that we are in the creature's domain and it has a huge "home town advantage!" Let me tell you why. Each time I saw the creature, it was clear that it was much calmer than I was. It was always a shocking experience for me, but the creature was usually not nearly as nervous about the encounter as I was! It seemed that it probably knew I was there, long before I got there, and it knew it was in no danger. It knew it had the upper hand and could move away at any time. They are incredibly fast and agile and can move quickly and easily through the thickest and most inhospitable terrain anyone can image! After satisfying its own curiosity, it would simply disappear again! A few times, if I was too close, it might rock back and forth in a threatening manner, before disappearing in the brush. Looking into its eyes is an experience I will never forget. It is difficult to explain, but you feel like the creature can see right through you. There is a connection that you can feel, but not explain.

LCS: You and your team actually go out into the field to investigate sightings and to search for bigfoot. How often do you do this? How many are on your team? Tell us about some of the tools and techniques that you employ during a research expedition.

TB: Sure Linda. I am truly blessed to have been able to assemble a core team of talented and committed individuals who demonstrate a passion and dedication that rivals mine. Along with this core team, we have an "extended family" of people all across this country who work alongside of us to share information and investigate sightings. From this constant flow of information, we are able to build a database about the habits and movements of these elusive creatures. We have been able to identify the most probable habitats and migrational movements throughout the nation. Once we get a report of a reliable and recent sighting, the core team heads out at a moment's notice to investigate. We always go prepared to attempt a live capture if it is possible. We have found lots of physical proof of the existence of this creature, but until we can show an actual creature to compare the evidence with, it doesn't mean much to the mainstream scientific community. We have lots of DNA evidence, but without an animal to compare the DNA with, it doesn't prove anything. You see, if you find DNA from a cat, and compare it to the data base DNA of a cat, you can say it is a cat. However, if you find DNA and have nothing to compare it to, you can't say what it is. The DNA evidence we have now, always comes back "unknown animal, nothing on record, possibly a primate." That is why nothing will prove the existence of this creature except a physical example. We need to bring back a creature to end this controversy. We have developed a great deal of respect for this animal over the past few decades and do not want to harm it. That is why we bring specialized equipment such as custom-made nets and bio darts to attempt a live capture.

LCS: What do you plan to do with the creature if you capture one?

TB: Linda, our plan is simple. We have a safe place to hold the creature, already designated, and the professionals lined up to care for it, you know, veterinarians and the like. We would want to allow the scientists and other professionals to study the creature and do their (non-intrusive) tests, such as hair and blood samples, X-rays, physical measurements, etc. We would plan on holding it for a MAXIMUM of ninety days. At the end of that time, we would want to transport it back to where we found it and release it back again.

LCS: Do you think that their discovery would further endanger the species, or that they would somehow be protected?

TB: We would hope that the information gained about the creature and it's habitat would be influential in establishing some regulations on securing its safety and the protection of some of its migrational routes. However, we do not make policy and it would be up to our government and the people to determine the next step.

LCS: What would you speculate is the history of this animal? Are they descendants of *Gigantopithecus? Homo erectus*/Neanderthal? Primate? Or are they something else entirely?

TB: We don't know enough about this creature to say for sure. That is one of the reasons we are willing to continue making the sacrifices we make and continue the hard work we have started. This is a quest to answer questions that cannot be answered today. I do, however, have my opinions. I personally believe this is a descendent of the yeti of the Himalayas. I believe it migrated here over the land bridge, near where the Aleutian Islands are today, the same way it is believed the native peoples of this continent got here. It is probably some type of primate and may even be distantly related to man, as are the great apes of Africa.

LCS: You believe these creatures are migratory. When do they migrate and why would they need to do this? Can you elaborate on this theory please?

TB: Our studies clearly show there is a definite migration pattern that takes place. As a matter of fact, I believe I was the first to coin the phrase "migratory creature" when discussing these creatures. We have seen that it migrates north to south and again south to north depending on time of year. I believe it follows the weather, much the same ways that birds migrate and for many of the same reasons. We have countless stories of creatures spotted in a specific location during a specific month and not seen again until that same month a year later. Just this past week, one of our guests on the *Bigfoot Live Radio Show,* (bigfootliveradioshow.com), shared a story of a creature that was investigated by both the Alabama and Florida State Departments of Fish and Game. This creature came up through a river bottom and killed some livestock at a small home farm in the area. Footprints and hair samples were found, but neither department was able to positively identify the creature responsible for the killings. Even the DNA collected came back, once again, as "non-human, unknown animal, nothing on file." This went on for three or four years, during the same month of each year. The family always knew when the creature was back, because they could smell it. They talked about a very peculiar and powerful odor that they noticed when the creature showed up. They eventually stopped raising livestock in this place, but the odor still comes back every year, during the same week or so of the same month. It continues to this day, over thirteen years later!

LCS: What can you speculate about the level of their intelligence?

TB: They are extremely intelligent. They know their habitat and how to utilize it better than any other animal I am aware of. They are very careful not to be seen, moving about almost entirely at night. The only thing that they don't seem to understand is that the only thing that gives them away is their footprints. I believe that the information we learn from finding these prints, will eventually give us the knowledge we will need to make a capture. The data that I have compiled over these past decades allows me to predict where and when they might be found at any given time. This, along with the small army of dedicated associates across this great country of ours, gives me and my team a distinct advantage towards a live capture, something I believe is imminent within the coming months.

LDS: What's the theory of why there's never been any bones recovered? And why haven't we found a dead body? Do you think that bigfoot bury their dead?

TB: I am always surprised to hear this question. The answer is really quite obvious to anyone who spends time out in nature. Nature has a way of keeping itself clean and healthy through an army of "sanitary workers." Scavengers of all kinds will clean a carcass down to bone in just a few short days. Others, like rodents and birds, will eat and scatter bones within a few weeks. And finally, the natural process of weather will finish the job very soon after that. There must be millions of bears in this country, but it is extremely rare to find a bear skeleton in the woods. Deer shed their antlers every year, but how often do people find a complete antler lying around in a field? When these types of items are so rarely found from common animals, how much rarer would it be to find something from an animal as scarce as this creature? Most wild animals seem to sense when they are nearing death and seek out and find a remote and quite place, as safe as possible from predators. If this creature does the same, the chances of finding the remains would be almost none. And, if a piece of bone did survive, and did not get buried by leaves and natural debris, how would anyone know what it was or think to pick it up? No, just common sense tells you that finding a skeleton would have to be much rarer than having an actual encounter with a living specimen!

LCS: Investigators have reportedly found samples of hair and scat that yielded samples of DNA, which did not match humans or any other known animals. Why is this still not accepted as proof?

TB: I think I may have already answered this question, but to make it clear, DNA is only a valuable tool when making comparisons with a known species. DNA is specific to a certain species of plant or animal, but without a standard for that species, DNA cannot be used to identify an unknown plant or animal. There have been studies done to compare specific DNA markers found in the many hair and scat samples found, that show common markers. This means that there is clearly some kind of unknown animal roaming this continent, but it does not yet say what that animal is. That is what Tom Biscardi and Searching for Bigfoot is going to do!

LCS: Can you tell us about your radio show, where we can find it, and what we might expect to hear and learn on your show?

TB: Our popular live broadcast is called the *Bigfoot Live Radio Show.* It can be found at bigfootliveradioshow.com, or you can link directly to it from our website: www.searchingforbigfoot.com. The show is on, live, every Wednesday evening at 7:00 p.m. PST. We generally have guests who are regular people who are surprised to have had some type of encounter with this creature. Those people have an opportunity to share their experience in an atmosphere where they will not be ridiculed or embarrassed. We strive to make the experience as comfortable and easy as possible for our guests. Unfortunately, many people who have a chance encounter are reluctant to discuss it for fear of ridicule, and may go many years without mentioning it to anyone. They feel very relieved to find out that they are not alone, and, in fact, are blessed to be among the few who have seen it. In other words, we share real stories, from real people, in an informative and relaxed atmosphere. It is entertaining and informative to both the people who know the creature exists and to the people who may just be curious about the subject.

LCS: Do you have any advice for the average guy who wants to go out in the forests to search for signs of bigfoot?

TB: Yes, I do have some suggestions. First of all be prepared; the forests can be a difficult and dangerous place. Make sure you are physically able and prepared before attempting an outing. Take a buddy along and make sure you have let people know where you are planning on going, what trails you will be using, and when you expect to be returning. Do your homework and make certain you bring all of the tools you might need to get in, out, and back safely. Make sure you know how to use them properly before you go. Know what natural dangers there are in the area you will be exploring. Know what animals, snakes, insects, and hazardous plants might be found there and how to avoid them. Bring a good camera and lots of water with you. The most important thing to remember is that it will require a lot of background knowledge and planning before you attempt it. It is always best to check in with your local forest service station and ask for their help and suggestions before planning any trip. Even a simple day may yield results if you plan it carefully and correctly. And if you are really serious about this, you might consider contacting me through our website. If we are coming near your area, we may be able to find room for you on an actual expedition.

LCS: Do you have anything to say to the skeptics and those who say, "That whole bigfoot thing is just a myth!"

TB: Sure, these people are in good company. They number with those people who said there was no such thing as a giant squid, a mountain gorilla, or any other of the hundreds of new species of animals discovered each year. But, I might also suggest they keep their eyes on me and my team, because they will soon be very shocked at what we plan on showing the world!

LCS: Any final thoughts you'd like to share in closing?
TB: If anyone out there has had an experience with this creature, don't hold it a secret. You are not crazy, it is really out there, and very soon, everyone will know about it. Give me a call and share your story and help me to find the final small key that will solve this mystery once and for all.

LCS: Thank you for taking the time to share with us Tom. Good luck with your research and please keep us posted.

TB: And thank you, Linda, for your interest and open approach to this subject. I am looking forward to having you share this project with me and my team. If you come up with any other questions, or if I can be of any help to you, please give me a call and we can discuss it.

Tom with his crew of sasquatch seekers (left to right), Steve Kulls, Jim Snell, Java Bob, T.J. Biscardi, Tom Biscardi, Brian Mazzola, and Father Ray Mahlmann.

Thomas Steenburg
June 2008

Thomas Steenburg

LCS: Hello Tom, can you tell us about yourself and how you became involved with the sasquatch phenomenon?

TS: My interest goes back to when I was a kid and first learned how to read. I remember having a keen interest in the Loch Ness monster after my folks ordered a large hard cover Reader's Digest book, which covered almost all that was known about our little blue planet at the time. In the section about the dinosaurs appeared two pages of those well-known black and white photos of Nessie. Some switch clicked on and I wanted to find out all I could about this and other mystery creatures reported about the world. I must have been a practical kid, as I knew at a very young age that I would never move to Scotland. Sasquatch being in my own back yard, so to speak, soon over took Nessie as interest number one. Sneaking into the local theater about ten times to watch, *The Legend of Boggy Creek* then solidified that interest.

LCS: How do you conduct your investigations in the field, what equipment, if any, do you use? What are your tools of the trade?

TS: I do not have the good fortune of having many of the fancy toys available to some of my colleges. Since I started in the late seventies everything has been at my own expense and I have never been a rich man. I do have all the necessary camping equipment and never go anywhere without my cameras, binoculars, tape recorder, sound amplifier, measuring tape, plaster of paris, note book, pencils, flash light, compass, maps, and of course, gun.

LCS: What is the most enjoyable aspect of this work for you?

TS: Being out in the mountains, and meeting fine people who feel that what they saw just must be told to somebody. Also, sharing a common interest with fellow researchers.

LCS: Many eyewitnesses mention the overwhelming feeling of being watched. Have you experienced this? And what are your thoughts on this feeling?

TS: When out in the mountains you are always being watched by something, sasquatch not withstanding! I have interviewed witnesses who suddenly have this feeling and look around to find the creature standing so still in the shadows that if they had not looked right at it they would have never known it was there. Of course, once the creature realizes it has been spotted, it leaves the area. I have had such a feeling a number of times, though it never resulted in a sighting of my own.

LCS: Bigfoot has been known to grunt, growl, scream, hoot, roar, whistle, sound like a woman talking, a baby crying, and has also been known to mimic our language. What are your thoughts on these vocalizations and their meanings? And do you think they communicate with each other in a language of their own?

TS: I think they can communicate with each other, but whether it can be described as a language remains to be seen. With most of these sound recordings collected, no animal was seen when the sounds were heard! Sounds in the forest are just that, sounds in the forest, unless the witness sees what was making them. Just a couple of years ago, Gerry Matthews and myself were investigating what has come to be known as the Chehalis sounds. These recordings were the same as other well-known, alleged sasquatch screams dating back to 1969, ones that I would have bet money on would be proven to be sasquatch in origin. Well, on April 5, 2006, we witnessed what was making them—a coyote! This discovery did not sit well with some in the bigfoot field. But facts are facts. As far as I know, the only sound recording made in which a sasquatch was observed in conjunction with the sounds at the time were the recordings over the police mike in Sgt. Kenny Cooper's patrol car. Cooper was a deputy sheriff on the Lummi Indian reserve just below the Canadian border. This happened in 1977.

LCS: In the Pacific Northwest, sasquatch have been spotted from Northern California, Oregon, Washington, and up into Canada. Do they migrate? Are they nomadic and travel depending on the season and the weather, or because of food sources? Or do they tend to stay and live in one area?

TS: Who knows? If I had the answer to that we would have ended this a long time ago. Reports do seem to come from the same locations in all seasons. But there are communities in which there have never been reports where suddenly there is a rash of them for a time, and then they stop as suddenly as they started—as though the creature hung around for a while and then moved on. I tend to think this occurs only when an individual has come close to a community for whatever reason. If the sasquatch moves around a lot in the backwoods, well, we just do not have enough information at this time to even render a guess.

LCS: If you were to come face to face with a sasquatch, what would you do? And what do you suggest others do? Stand perfectly still? Run away as fast as you can? Try to communicate with it?

TS: Pictures! Video! Shoot it, if you think you can bring it down!

LCS: If a person were to capture a sasquatch, or find a dead one, what would be the correct thing to do?

TS: I cannot imagine anybody capturing a sasquatch, not alive anyway. If someone were to come across the remains of a dead one out in the bush, first thing is to take as many photos or video as you can of the body, from all angles. Make detailed notes of the site, and of the cadaver. Then if the remains are too heavy for removal, take the head and the hands and the feet. You will feel a bit foolish if you drag someone out there later and the body is gone!

LCS: In your years of research, what are some of the most common misconceptions people have about these animals?

TS: I think those who think this creature is more of a primitive human than ape are way off the mark, as the witness evidence does not, as far as I'm concerned, support this hypothesis. If one closely listens to what the subject was doing and how it was behaving when spotted, it seems to me that we are dealing with an ape not a human. Also, too much attention is taken up by things like twisted trees, nests, sounds, wood knocking, rock piles and many other such things which many assume a sasquatch is responsible for when there is no evidence of this.

LCS: Do you have any advice for the average person/researcher who wants to go out in the bush to explore and look for signs of a sasquatch?

TS: The best advice I can give is do not go anywhere without a camera in your hands. If it happens, you will only have a few seconds to react, and if you fail to get a photo or video you will simply be another person claiming to have seen the sasquatch.

LCS: If you have one piece of advice for future generations of sasquatch researchers, what would it be?

TS: Do not put the desire to find the truth of this ongoing question ahead of other priorities, i.e., family.

LCS: In closing, is there anything you want the world to know about these intriguing creatures? Any final thoughts?

TS: I would just suggest to people who do not think that anything is out there, or who do not care, to at least keep an open mind to the possibility. Science has been wrong before and they may be wrong again. And if the facts eventually convince us that this creature does not, and never did, exist, then we can say we did our part to catalog a very important part of North American folklore. But if the day comes when hard physical evidence is produced, then we can all be assured that true wilderness is still out there, which we know little about, and that will sure be a kick in our collective apathy.

LCS: Thank you for taking the time to share your knowledge with us Tom. Keep up the good work!

TS: You are most welcome. Keep your eyes open.

Thomas Steenburg provides highly detailed accounts of personally gathered eyewitness sightings and related incidents in his books (Hancock House Publishers). Tom leaves no stone unturned in his interviews and investigations.

JOHN GREEN
June 2008

John Green

LCS: Hello John, it's an honor, a privilege, and a pleasure to interview you. Can you tell us about yourself and your background? And when and how you became interested in the sasquatch phenomenon?

JG: I am a newspaperman, now long retired. Tales of sasquatch as giant, hair-covered, semi-humans have been in print in British Columbia since before I was born, but from a single source, and I did not take them seriously until 1957 when I heard a firsthand report of huge human-like footprints from someone I already knew and respected. On investigation it became clear that many apparently-credible people tell of seeing such creatures, and my interest was aroused to the point that in 1958, when the first report of bigfoot came out of California I went there to look into it. There I saw fifteen-inch footprints sinking almost an inch into hard-packed sand where my own prints scarcely showed, and for the past half century I have been trying to establish how those prints were made.

LCS: Of the thousands of sasquatch related incidents you have gathered, what are the most compelling pieces of evidence that convince you that these creatures are real?

JG: To me, the three most compelling individual pieces of evidence are the movie of such a creature taken in 1967 in California, the excavation left where a witness told of seeing a sasquatch dig out and eat hibernating rodents in Oregon that same year, and body and heel imprint cast recently in Washington.

LCS: Certain audio recordings have been made of "bigfoot language." It sounds almost like gibberish. Do you feel that these sounds could be an actual language?

JG: I know of no recordings of sounds a sasquatch was seen to be making, and of the wide variety that are attributed to them, none could possibly be considered a language.

LCS: Do you believe the foul smell commonly reported is just the natural way they smell? Or is it a pheromone they release only when they feel threatened? And do you think they can control this release?

JG: Since no smell is reported in the majority of instances where the witness was in a position to notice one, it seems probable that a controlled release is involved.

LCS: Many people have mentioned an overwhelming feeling of being watched in connection with a bigfoot incident or encounter. What are your thoughts of this experience? And have you experienced this feeling yourself?

JG: I have never experienced any feeling of being watched. I don't doubt that other people have, but the cause remains to be established.

LCS: Bigfoot has been known to leave gifts in exchange for food left for them. What do you know about this amazing habit of theirs?

JG: There are just a few stories of this sort, and while it may happen I wouldn't call it a habit.

LCS: Do you think sasquatch is bothered by the cold and snow? What is your opinion of what they do and where they go in the winter?

JG: I have no idea of the extent to which sasquatch are bothered by weather, but I do think they must hole up in winter, since there are relatively few tracks found in snow.

LCS: What do you think about the theories some people have that these creatures have some sort of mystical powers, are shape shifters, exist in a separate plane of existence, or have something to do with UFOs?

JG: I see no sense in trying to explain an unproven phenomenon by attributing it to some other unproven phenomenon.

LCS: What is the oldest sasquatch story that you know of in the Pacific Northwest that you believe to be true? And what is your favorite story, recent or classic?

JG: There were several sighting reports on Vancouver Island between 1900 and 1905 that were investigated before my time and considered to be true. I don't know about "favorite," but Glen Thomas' story of watching the sasquatch dig out the hibernating rodents, and of course Roger Patterson and Bob Gimlin's account of their experience are very impressive and backed by physical evidence.

LCS: What is the most burning question in your mind that remains unanswered? What do you want to learn about them above all else?

JG: Nothing is burning. I would like to live to learn the results of a physical examination of one of these creatures, but my odds are pretty short now.

LCS: If you have one piece of advice for future generations of sasquatch researchers, what would it be?

JG: Do whatever gives you satisfaction, so that you will be more likely to stick with it, and keep in mind that distant fields are not necessarily greener. You will be able to do more if you can do it near home.

LCS: You are considered to be "the legend among us;" your books are the cornerstone of sasquatch research. I would be honored if you would list your published books here for our reference.

JG: *On the Track of the Sasquatch,* 1968
Year of the Sasquatch, 1970
The Sasquatch File, 1973
Sasquatch, the Apes Among Us, 1978

As to revisions, basically of *On the Track,* 1980 and 2004. Currently only *Apes Among Us* and the revision *The Best of Sasquatch Bigfoot* are in print.

LCS: What do you think is the biggest mistake made by researchers today?

JG: Assuming that things that may well be attributed to the sasquatch are established fact, and forgetting that nothing short of a body is going to provide indisputable proof.

LCS: Thank you for taking the time to do this interview, John. It has been a pleasure and an honor. You are truly a legend.

In this 2004 work, John Green presents the latest scientific developments in the sasquatch arena, together with the best sasquatch accounts and other information he has collected in over forty years of research.

This definitive work by John Green was first published in 1978 by Cheam Publishing Limited, Agassiz, British Columbia. It was republished in 1981, and again republished in 2006 by Hancock House Publishers. It remains the most comprehensive and highly acclaimed work in the field of sasquatch studies.

Dmitri Bayanov
June 2008

Dmitri Bayanov

LCS: Hello, Dmitri, it's an honor to interview you. Could you tell us about yourself and how you became interested in the bigfoot phenomenon?

DB: I am a student of a new discipline that I named hominology, and I am a "freelance philosopher." I received a college education, but have no scholarly degrees because my scientific interests and works have been in the subjects ignored or rejected by mainstream science. I was born in Moscow, Russia, in 1932. I am married and have a son and two granddaughters.

As a boy, I was very interested in animals, and visits to the zoo excited me much more than visits to the cinema (not any longer, for I hate seeing animals in cages).

On June 22, 1941, Nazi Germany attacked the Soviet Union, and history's most terrible war started. Moscow was repeatedly bombed, and my father took the family (mother, sister, and me) to

Tajikistan (then part of the Soviet Union), far away from the front. He was soon called-up into the army, and we stayed in Tajikistan until the end of the war in 1945, when we returned to Moscow.

While in Tajikistan, we lived in a small town called Shakhrinau, and it was there and then that I heard for the first time of "hairy wild men" living in the mountains, but could hardly believe the rumor. I recalled it decades later, when I revisited Tajikistan in 1982 on a hominological expedition, as described in one of my books.

At the time, I continued to entertain great interest in animals and dreamed to become a famous zoologist, like the eminent German naturalist Alfred Brehm (1829–84), whose big, well-illustrated volume, translated into Russian under the title *The Life of Animals*, I read and re-read all the time. One trait of my character was high curiosity, and later, in Moscow, schoolmates used to make fun of it, teasing me with the phrase of a fictional character who liked to ask improbable questions, like this one: "What if from an egg an elephant is hatched?"

I guess I was eleven, and still in Tajikistan, when a strange and hardly believable thing happened to me. One late evening, in the fall, during one of my "nature study" outings, I noticed a flight of goldfinches descend in a tree for a night rest. Suddenly, a daring thought struck my mind: Why not catch one of these lovely birdies? It was an "if" thought: if I succeed, if I catch one, my dreams will come true. I'll become like Alfred Brehm, or get notable in some other way.

Very cautiously, very slowly and quietly, mimicking a hunting feline, I started climbing up the tree. It was semi-darkness, and I discerned one bird, closest to where I was in the tree. I well remember that my hand trembled as I stretched it out toward the bird. I snatched it; it let out a piercing call of distress, and the whole flock instantly and noisily took off from the tree.

Carefully holding the bird in my hand, I managed to climb down, returned home, showed my trophy to mother and told her why I got it. Then I released the captive.

Well, I am not like Alfred Brehm, but some success, in a field beyond my imagination at the time, has been achieved.

That episode was fine and marvelous, and it happened during the daily manslaughter of war in the west of the country, on a thousands-mile front from north to south. Tens of millions perished in it, and there was hardly a family in the country that did not have a member killed or maimed. One of my uncles, the best-loved one, was killed in the Stalingrad battle. I always feel and remember that I, and my dear ones, are alive because he, and millions like him, gave their lives. There would have also been little chance for me to survive if I had been born a few years earlier and took part in the battles.

Some people today imagine that Nazism in Germany and "communism" in Russia (it was not at all what Marx had termed communism) were equally bad. That is wrong. The Soviet system was an inadequate and in some ways evil and criminal execution of fine and humane ideals ("the road to hell is paved with good intentions"), while German Nazism was a highly efficient execution of utterly evil and criminal ideas. Whatever good was in the Soviet Union, especially in culture, medicine, and education, it was due to the ideals of justice and humanism, proclaimed in its ideology. Whatever evil was in Nazi Germany, it was due to its evil ideology. This shows the supreme importance of ideas that people follow, and consequently, of people's humanistic education and enlightenment. On the whole, the Second World War, in which Russians and Americans were allies, was a virtual Armageddon between the forces of relative humanity, on one side, and those of absolute inhumanity, on the other. That is, if the events are taken and viewed on a historic and global scale. As to individuals involved in that global beastly disaster, there were, as usual, upright and humane people on both sides.

I mention here these historic events because their impact had much to do with my pursuit of science, including hominology in later years. But to finish with my experiences in Tajikistan, I should add that I learned there full-well the feel of starvation caused by war deprivation. It's a most humiliating condition, turning a human being into an ever-hungry animal, unable to yearn for anything but grub. One winter, our food situation was so bad that I missed one school year, unable to do homework for loss of memory caused by starvation.

On the positive side is the memory of the delicious taste of canned margarine we began to get in 1943 or 44 under the American Lend-Lease program of aid to the Soviet Union, as U.S. ally in the war. To this day I can't help feeling that margarine tasted better than any butter I ever ate in my life.

Another marvel was U.S. Studebaker trucks and Willys jeeps

that I saw for the first time. Once in the mountains, I gaped with admiration at a Willys jeep that with unusual ease dashed up a very steep road. That was in Tajikistan, a remote corner of the vast Soviet Union, which covered one sixth of the whole world land area. Think therefore, what a stupendous amount of technological and food aid that was being sent by America to its ally to reach and cover the country's still huge unoccupied territory. While America was donating so generously food and technology, the Soviet people were sacrificing for the sake of victory a sea of blood.

So as a youth and young man back in Moscow, I gave much thought to what happened to mankind as a result of a second and much nastier world war; what happened to the German people under Hitler, and the Soviet people under Stalin. From zoology my interests shifted to philosophy, sociology, and anthropology. It was due to these interests that in 1964 I met Professor Boris Porshnev, who acquainted me with the problem of so-called relict hominoids, and that was the start of my hominological investigation, including the bigfoot phenomenon.

LCS: What are the Russian names for what we refer to as bigfoot or sasquatch?

DB: These names are divided into old and new. Old names are many and retained in folklore, in everyday speech (in sayings and proverbs), and dictionaries of the Russian language. The most common among them is the name *Leshy,* derived from the word *les* (wood, forest). Russian–English dictionaries translate Leshy as wood goblin. According to folklorists and ethnologists, Leshy is a figment of people's superstition and has no basis in reality. Hominologists, on the contrary, have enough evidence and arguments that Leshy is an old and original Russian name for a "forest wild man."

The mention of wood or forest is present in numerous other ethnic names of our hero. As to superstitions, mythology and magical beliefs connected with this subject, they are based on the real extraordinary physical and psychological powers of these hominids.

The new name that appeared without any connection with Leshy, and before the birth of hominology, is *snezhny chelovek* (snowman). It's a shortened translation of "abominable snowman" that entered the mass media vocabulary in the middle of the 20th century, especially as a result of the British 1954 yeti expedition in the Himalayas.

There is still no generally accepted scientific term for Leshy and bigfoot in the flesh. My handy working term is *homin.* Hominology is the science of homins.

LCS: Strange "wild men" are depicted in Russian drawings, paintings, sculptures, and engravings. How far back can you trace the history of these wild men?

DB: Images of "wild men" are to be found all over the world. We see them in prehistoric petroglyphs, in the art of the ancient world and the Middle Ages, and in recent and modern folk art. Their telltale signs are depiction of hirsute bodies, usual lack of clothes, and certain features of the face and figure that distinguish these images from those of modern humans. This is an important category of evidence and source of information for hominology.

LCS: What do you speculate is the nature of the Russian forestman and bigfoot? Are they descendants of *Gigantopithecus,* or entirely something else?

DB: The exact biological, i.e., genetic and taxonomic, nature of these bipedal primates is unknown. I strongly believe they are not descendants of *Gigantopithecus.* I believe they are as close to modern man, *Homo sapiens,* as are Neanderthals and/or *Homo erectus.* It is noteworthy that in three cases known to me (one in Russia and two in America), witnesses among common people said that these "wild men" look like Neanderthals (whose pictures were known to them from books).

LCS: Can you tell us how the Russian forestman differs from the North American sasquatch in relation to size, footprints, vocalization, and appearance? And do you believe they are of the same or a different species?

DB: Judging by available data, I don't think our forestman, Leshy, is noticeably different from your forestman, bigfoot. I think that if they are different, then at the level of races, i.e., subspecies.

LCS: How does the Russian forestman differ from the North American sasquatch in regard to its habits and interactions with humans, dogs, or other animals?

DB: Here too, I see no difference. Even in their habit of braiding the manes of horses they are alike.

LCS: Tell us about some of the techniques that you have employed during research expeditions.

DB: The main technique during our research expeditions is the correct way of approaching and interviewing witnesses. As a rule, people are not eager to share their experiences of encountering or interacting with forestmen. And when they agree to talk, it is essential to ask the right questions and record the answers correctly. For this purpose we have special instructions and list of questions. Coming to a village in a homin habitat, we introduce ourselves not as hominologists but as zoologists and ecologists researching local wildlife. Thus, from questions about elks, wolves, bears, etc., it's easier to pass on to the question about a creature that is like a bear, but walks erect on two feet.

As to fieldwork, searching for homin footprints and their photographs and casting are the usual practice. Knowing that many witness accounts have been received from shepherds and farmers, I dreamed of creating and using "mobile animal farms" of our own, placed in promising areas, hoping that our farm animals (horses, goats, rabbits, etc.) would attract the homins. But to be realized, such plans require lots of funding, and we have always had none.

Our late Leningrad (now St.Petersburg) colleague, zoologist Rostislav Danov, who said he had tasted the flesh of nearly all animals that died in the zoo, intended to adopt the exact lifestyle of a "wild man" in case he happened to see fresh tracks of the creature. In this way he hoped to make friends with the forest people and begin directly observing them. Rostislav made a living by catching and selling poisonous snakes (for medicinal purposes) in the deserts of Central Asia, and before he had a chance to test his technique, he died of multiple complications following a snake bite.

The best technique that is the most fruitful so far, has appeared spontaneously in America among people, women in particular, who befriend and observe bigfoot on their property. Your grandmother was a pioneer in this regard. In truth, you don't find bigfoot; bigfoot finds you.

LCS: Are there some areas or regions of Russia where the forestman has been spotted more than other areas or regions? And what is their preferred habitat?

DB: In the beginning, following the yeti expeditions in the Himalayas, we believed that the forestmen could only be met in the Pamir Mountains of Tajikistan, closest to the Himalayas. Then sighting reports began coming from the Caucasus, much closer to us in Moscow. Now we have sighting accounts from all over Russia, including the Moscow Region. I conclude that in some areas, where the pressure of civilization on wildlife is the least, extant hominids enjoy permanent habitats; and in areas where pressure is high, they probably lead a Gypsy-like vagabond life, appearing here and there from time to time.

LCS: What is the one thing you want to learn and know about these creatures (both the Russian forestman and/or bigfoot)?

DB: The one thing I dearly want to learn is their linguistic ability. This will show how closely they are related to us, and if we have a means of mutual understanding. From almost all historic and modern evidence it is inferred that they have no speech ability. On this account, and for certain theoretical reasons, my teacher of hominology, Boris Porshnev, maintained that only modern man, *Homo sapiens,* developed the gift of speech, which is his exclusive possession and the cause of origin. I believed this theory and spread it. True, there were a few exceptions in the evidence, but they were either ignored or explained away as inaccuracy in witness accounts. Now, after the 2002 book, *50 Years with Bigfoot,* by Mary Green and Janice Carter, I accept the possibility of verbal language, at least in some non-sapiens hominids. This means that the Porshnev theory may have to be seriously revised. Igor Bourtsev holds strongly that some homins, thanks to their high mimicking talents, "picked up" language from humans. And I've always supposed that, even if Porshnev was right and speech is modern man's invention, a young homin could be taught to speak. Anyhow, the question of their linguistic ability is the most pressing and educational.

LCS: If you could have the attention of the whole world, what is the one thing or message you would want to convey about these intriguing creatures (both the Russian forestman and/or bigfoot)?

DB: I would want to convey three wishes: May these nature people teach us to live in peace and harmony with Mother Nature. Seeing in the dark, may they help us see the light of reason. May they help us learn to be *Homo sapiens*.

LCS: Thank you, Dmitri, for this interview. I greatly appreciate your thoughts and insights.

DB: You are welcome, Linda. Good Luck with your book.

Dmitri Bayanov was the first researcher/author to provide comprehensive information on both the Russian snowman and the Patterson/Gimlin film. His books (Crypto Logos Publishers, Moscow, Russia) are highly authoritative works on these subjects.

Peter Byrne
July 2008

Peter Byrne

LCS: Hello Peter. For the few readers that are not familiar with who you are, can you tell us your background and how you became interested in the bigfoot phenomenon?

PB: For me, bigfoot, the great search, as we called it, began in a hotel in Katmandu, in late December 1959, where I and my brother were taking a break after three years of yeti hunting in the Himalayas. Among the mail waiting for us was a cable from the great Texan explorer and entrepreneur, Tom Slick (our expedition's sponsor) asking us to come to the US to design and run a project to investigate what he called the bigfoot mystery. Three weeks later we were in Texas, sitting down with Tom and studying maps of the Pacific Northwest (and finding the area, to our astonishment, to be five times larger than the Nepal Himalayas), and a week after that we headed north. That was the beginning.

LCS: Can you tell us something of your lifelong interest in wildlife conservation and your past and present projects taking place in the small Himalayan country of Nepal?

PB: My interest in wildlife conservation came out of a passion for hunting, which same began for me in the green hills of Ireland, where I was raised. I hunted there and later in India as an amateur, turning professional in 1953. Eventually, like many hunters, I developed a great respect for wildlife, especially for the big cats, and mainly because of this gave it up completely (in 1970) and turned to conservation. I now work in conservation for six months of every year in the country of Nepal (spending the remaining six months in bigfoot phenomenon research in the Pacific Northwest). The best source of information on my conservation work can be found under the website of the society with which I have worked for some forty years, which is: www. internationalwildlife.org.

LCS: Your search for the elusive bigfoot has taken you on adventures to many states and countries? Do you have a harrowing story of an encounter, or close call to relate?

PB: Many countries? No, only in western Canada (British Columbia), and in the U.S. only in the Pacific Northwest. As to one experience, not necessarily harrowing—one night, in a remote area of the Cascade ranges of central Oregon, something walked around my camp (in an area where there had been two credible sightings within the previous forty-eight hours) cracking large sticks for more than an hour. The ground within the dark wall of trees that surrounded my campsite had a thick layer of conifer debris; there were also several heavy, crunching footsteps. Was it a bigfoot? No North American animal is capable of doing something like this; to break large sticks like this, hands, powerful hands, would have to be used. Because no one knew I was going to be camping in that area, and because of its isolation, it is doubtful that the sounds were of human origin.

LCS: Bigfoot has been known to grunt, growl, scream, howl, roar, whistle, sound like a woman talking, a baby crying, and has been known to mimic our language. What are your thoughts on some of these vocalizations and their meanings? And do you think they communicate with each other in a language of their own?

PB: It is my opinion that, after some forty-seven years of study and research on the phenomenon, of which seventeen were full-time professional research programs that cost millions of dollars, with the possible expectation of whistling, the creatures are silent and that all the reports of roars and screams, etc., are without any basis in reality. As to whether they communicate with each other—it's possible. But if and how they do, we have not a clue at this juncture.

LCS: Many people have mentioned an overwhelming feeling of being watched in connection with a bigfoot incident or encounter. What are your thoughts of this experience? And have you experienced this feeling yourself?

PB: I have people tell me about this kind of experience, but in the thousands of hours I have spent in the field in BC and in the Pacific Northwest, I have never experienced anything like this, nor have any of my companions. In the dark and often gloomy forests of the northwest, especially at night, it is not difficult for people to image they are being watched by some unknown entity.

LCS: Different witnesses report slightly different physical appearances of bigfoot. Are there different breeds or races of bigfoot out there? Perhaps different cultures in different states and countries?

PB: It is my opinion that there is only one species, the large one known as the sasquatch. The main differences of description that I have encountered in my research are in the areas of color and height, the color ranging from dark brown to black and, in one case, white, and the heights of adults varying from six to eight feet.

LCS: Through your studies, research, and adventures in the field, what in your opinion is the best piece of evidence for the existence of these animals?

PB: I refrain from calling them animals because of the very significant physical attributes which strongly suggest they are hominids, including upright stance, bipedal locomotion, white coloration in the eye, and, according to credible eyewitnesses, something very close to a human face. The best evidence? One, their historical background, which I have traced back in old newspapers and magazines and letters written by early settlers and Native American lore to 1775. Two, the findings of large, human-like, five-toed footprints in remote areas. Three, the 1967 Patterson–Gimlin film footage. And, four, the very many eyewitness reports of encounters, supplied by mature, intelligent, highly credible persons, among whom, in my files, are senior government officials, including policemen and policewomen, engineers, surveyors, inspectors, and even high-ranking members of both state and federal judicial systems.

LCS: Being a former big game hunter in the past, do you advocate the shooting of a bigfoot? What are your thoughts on the killing of one of these creatures?

PB: The shooting of one? Absolutely not. People who senselessly promote the killing of one as a means of proving they exist are advocating nothing less than murder. These are my feelings, for the insensate and stupid people—the very few, thank heavens—whose bloody slogan is, kill it to prove it, which means nothing less than slaughtering a member of a rare and incredibly unique species, of which they may be but a few left.

LCS: What do you think you would do if you came face to face with one? How would you react?

PB: Hopefully I will react as I have done many times in my hunting days in confrontations with large, potentially dangerous, wild animals—calmly and thoughtfully, taking advantage of every single second of the encounter for observation and study and, with the "weaponry" without which I never leave home—digital still and video equipment—record the incident to the fullest extent that the circumstances would allow.

LCS: Do you think that their discovery would further endanger the species, or that they could somehow be protected and possibly helped?

PB: Would they be endangered by their (scientifically accepted) discovery? Not really. They have a huge range of protective habitat, Northern California to Alaska, literally thousands of square miles, which same, as an example of its ability to successfully hide them, contains fifty-four planes missing and undiscovered since WWII. As to protection, after discovery, there is little doubt that the governments of the both the U.S. and Canada would immediately enact inflexible laws for their total protection as a rare and endangered species, to be all the more stringent if they are recognized as a hominid form.

LCS: What is the most burning question in your mind that remains unanswered? What do you want to learn about them above all else?

PB: The burning and unanswered question? It probably is, who are they? In addition to that, of course, are all the other questions, not the least of which are: their intelligence, their natural skills, their social structure, their means of communication with each across a vast territory and the reason for their innate wariness of man, something that unfortunately (or fortunately?) persuades them against making contact with us.

LCS: What do you think is the biggest mistake made by researchers today?

PB: Believing ridiculous theories like those that tout the creatures as having extra-sensory and extra-dimensional capabilities, including, e.g., the potential to disappear at will. And also wasting time searching for them in places where there is no credible record of habitat, by which I mean all other states outside of western Canada and the Pacific Northwest.

LCS: Do you have any advice for the average guy who wants to go out in the bush to explore and look for signs of a bigfoot?

PB: Peruse and study the records of credible sightings and footprint finds, on a Geo Time Pattern (a pattern of place and time) basis, and use that to concentrate field work in the areas and at the times the GT pattern suggests. Use affordable technology, such as motion sensor cameras (which I use in my conservation work in Nepal, very successfully, for counting tigers and other shy, nocturnal animals) and night vision devices and digital cameras.

LCS: Is there any one thing that you wish the world to know? Any final thoughts?

PB: Being the only investigator in the bigfoot field who has spent a significant period (seventeen years) in full-time professional research on the phenomenon, I feel that this allows me, right or wrong, a certain seniority of opinion. On the basis of my extended research and field work—now going back some forty-eight years—and on my background as a hunter and naturalist, I would like to state that I truly believe that we do have a large, bipedal hominid living in the great forests of western Canada and the Pacific Northwest. I also believe that, in spite of (unfounded) arguments to the contrary, there are obviously not very many of them surviving. And lastly, I feel that, if only because of this latter, it is imperative that any future major approach to a find must be passive and must be planned with extreme circumspection, the foundation of which must be respect and esteem for the unique individuality of the species, with the uppermost thought being that, apart from what unknowns we might have in the depths of the ocean and outer space, the scientifically accepted discovery and identification of these extraordinary hominids will truly represent one of the greatest discoveries of all time to be made on our little blue and green planet—one of enormous scientific importance to man.

LCS: Thank you Sir. It has been a privilege and an honor. Best wishes to you in all your travels, adventures, and endeavors!

PB: I thank you for giving me the opportunity of voicing a few thoughts.

Peter Byrne's remarkable book, published in 1975 (Acropolis Books Ltd.,Washington, D.C., edition) and 1976 (Pocket Books, New York, edition) was the first major comprehensive publication on the sasquatch/bigfoot issue. It detailed current knowledge on the creature to that time and is now one of the great classics in the sasquatch/bigfoot field.

Highly familiar to many is Peter Byrne's Bigfoot Research Project logo. He directed the project during three phases that spanned 26 years.

The Bigfoot Information Center in The Dalles, Oregon (1970s). Peter Byrne arranged for the provision of this facility—the first of its kind to be provided. Through Peter's efforts, we had a central point for collecting information, effectively up to the current internet based organizations.

Chapter 5

From the Record

Every individual matters, human and non-human alike. Every individual has a role to play.

—Jane Goodall

NOTE: The following are selected short stories and quotations from various sources. Some have been either "handed down" or provided without any firm references. In both cases, we have no way to confirm or verify the information. As such, these stories are provided more for entertainment than factual findings. Others are, or appear to be, highly credible, and we have no reason to doubt the source. Whatever the case, they all serve to provide further insights into the bigfoot phenomenon.

The Chuckanut Mountains, 1470

Whatcom County, Washington: Indian accounts of early Spanish explorers in the area claim that bigfoot is the descendant of huge, bearskin-wearing, stick-toting slaves called "Stick People," who were used by Cortez to move the personal treasures of Axaquatl (Aztec ruler of the 1470s) and Montezuma, to a hideaway in the Chuckanut Mountains (outside of Bellingham, Washington) 450 years ago. The legend goes that the Stick People's descendants still guard the treasure, awaiting the return of their god, Cortez. High in the Chuckanut Mountains are images of reclining women carved in relief in the rock, apparently marking the front of a Spanish fort. (Source: International Bigfoot Society.)

Sasquatch Remains, 1833

Have sasquatch remains ever been found? Probably several. Have they been recognized for what they are? Almost certainly not. One such account comes from page 151 of Dr. Karl Shuker's book, *The Unexplained*: "In the year of 1833: A twelve-foot-tall skeleton was uncovered at Lompock Rancho, California, by soldiers digging a pit for a powder magazine. The skull had a double row of teeth." Also found were stone axes, carved shells, and porphyry blocks bearing symbols. The skeleton was reburied and "lost" when the locals started venerating it. (Source: International Bigfoot Society.)

Monkey-like Wild Man, 1904

In 1904, residents of an entire Indian village at Bishop's Cove, Vancouver Island, BC, abandoned their homes in a state of complete terror. Terror, because a "monkey-like wild man" took to spending nights howling on the beach in an unearthly fashion. The creature, which came out onto the beach at night to dig clams and howl, stood about five feet tall and was covered with long hair. When the steamship *Capilano* pulled into Bishop's Cove, the village inhabitants put off from the shore in canoes and clamored on board the ship in a panic. The Indians said they had tried to shoot the creature but failed, which added to their superstitious fears. (Source: *The Daily Province,* Vancouver, BC, Canada, 1904.)

Two Hunters Murdered, 1927

A report on file from 1927 tells of two hunters from the Quinault, Washington area who were found dead, their rifles twisted and distorted, and all their bones crushed and broken, as though they had been repeatedly smashed against the ground. (Source: Researcher Fred Bradshaw.)

Followed by a Bigfoot, 1957

In 1957, Charlotte S. lived near Malot, Washington (southwest of Omak). She walked one and a half miles home from school, weather permitting, along the paved Highway 97. There was a bus that she rode in the morning, but she liked the walk home. For a month, in May, she noticed several times that something was following her, as she could hear the gravel crunching on the bank above her. When she would arrive home, her collie/lab mix dog would be straining on its harness to get loose to get at the "something." One day she finally saw IT.

Charlotte was at home the day it came out of the woods to finally show itself; it stood only fifty feet away, with the sun at its back. It was about seven feet tall, dark colored; its head was flat on the top, there were tufts of hair on the sides of the face, and it had tiny ears. It was apparently attracted to the barn where there were goats and rabbits that had been disturbed at other times. "The dog spotted him and wanted to get loose, and the creature knew what I said when I swore at him," she said. Charlotte was fifteen years old and not afraid of anything, so she stood her ground and did not retreat. The creature did, and ran back into the brush. (Source: International Bigfoot Society.)

Knock on the Door, 1959

A story printed in the *Humboldt Times* of Northern California stated the following: "At 1:00 a.m. on a moonlit, early fall night in 1959, Lawrence Omeg heard a knock on the door of his shack. When he opened the door, a tall bigfoot was standing there. Just standing there looking at him. Not knowing what else to do, Lawrence handed the creature a candy bar and closed the door." (Source: *The Track Record.*)

Uncle Wendel's Encounter, 1963

Ralph Munson has an old tale concerning his Uncle Wendel, who was driving a truck for the Bend–Portland line in 1963. He was driving a blunt nose cab-over truck on the east slope of Mt. Hood, about fifteen to twenty-five miles east of the crest, and headed east. In the evening, it not being quite dark yet, he stopped when something jumped out from the woods and into the middle of the road; a deer he thought at first. Wendel said that he did not drink, and that an animal that looked like a gorilla, or more like an orangutan, came up to the truck and first looked through the front window, and then through the side window. Then it grabbed the cab of the truck and started to shake it. Wendel honked the horn and flashed the lights, and the thing ran away up a sheer bank that was very brushy. (Source: *The Track Record.*)

Rampaging Bigfoot, 1965

An anonymous lady said many years ago as a girl, that a bigfoot chased her and her companions down a hill at Raven's Roost, in Chinook Pass, Washington. In the same area in 1965, some kids were having a high-school party, and one of them shot at a bigfoot in the dark. The bigfoot apparently objected, and tore the boy apart. He was later found in a pool of his own blood, his ribcage crushed. (Source: International Bigfoot Society.)

Soldier Spots Bigfoot, 1967

Fort Lewis, Washington: At dawn, in the summer of 1967, a soldier was on patrol duty—at the time and the sun had just barely come up. Suddenly, a hundred yards off, a dark brownish, hunch-backed bigfoot, walked past the compound entrance on a dirt road, just "minding its own business." (Source: anonymous soldier.)

Truck Driver Scare, 1968

A Colville, Washington, Indian male, a fifty-year-old big-rig truck driver, reported an incident that occurred in 1968, in the early morning while he was taking on a load of water—filling a large tank on his truck from a stream. A pump was set up in the stream and while the tank was filling, he relaxed behind the steering wheel in the cab of the truck. There was suddenly a knock on the passenger side window and a huge *S'cwene'y'ti* (sasquatch) face was pressed against the glass. Its nose was flattened against the window; its hair appeared white.

Considering the height from ground to window, it would have to have been about eight feet tall, or more if stooping or bent over. The driver immediately started the engine and accelerated at the highest rate of speed possible, breaking the attached water hoses as he departed the area.

This same fellow reports that on another occasion, he saw a

S'cwene'y'ti on a hillside quite a distance away. He is positive that it was not a bear or other animal because of its bi-pedal locomotion and upright posture. (Source: Dr. Ed Fusch.)

Bigfoot Steals Deer, 1969

"Meat eaters they appear to be!" was the headline printed in a Skamania, Washington, newspaper in 1969, followed by this story:

> There were two good-ol' boys that had been out poaching this spring. They shot a deer up behind the old town of North Bonneville, or one of them had, and he went down to the local tavern and got his buddy to help him get the deer out. Arriving back at the scene, they found a sasquatch with this deer.
>
> These guys were kind of rounders, characters, but they were very honest people, and I know their story was the truth. They saw this bigfoot with the deer over its shoulder, and they both had rifles, but they took one look at what they called a "great big ape," changed their minds and took off. They got maybe a half-mile away, and one of 'em said, "Peter, we got rifles, lets go back there and get our deer." Well, they got back there all right, and found a lot of pieces of bone and meat scattered around, and quite a bit of blood. They looked at each other, and decided to just go back down to the tavern and call it a night. (Source: *Skamania County Pioneer*, March 25, 1969)

Brender's Pig Farm, 1970s

Brender's Pig Farm on Icicle Creek Road, four miles south of Leavenworth, Washington: Ten young fellows in the early 1970s were drinking on a Friday night at the farm. They had some heavy weapons handy; one had a Weatherby "elephant" gun. At 1:30 a.m. they heard a big noise in the barnyard. Thinking it was a bear, they had a competition to get to the door and shoot the bear, which was probably going after the pigs.

Instead of a bear, they found a huge thing with a 150-pound pig in its mouth, biting the neck. All started firing guns, and there were terrible sounds. One fellow knew he hit the seven-foot-tall, wide shouldered thing at seventy-five feet, and it was screaming bloody murder. It then dropped the pig from its mouth and began to drag its body away, all the while with the ten guys shooting at it. The next morning, they followed the blood trail on horseback for fifteen miles before losing the it. (Source: Researcher R.L.)

Canadian Soldiers Run into Bigfoot, 1970

Before his second tour in Vietnam, soldier Steve Bray was commissioned to Canada. "I cross trained with Canadian soldiers in a special mission. While there, I was told by one of the Canadian soldiers of an incident where his platoon literally ran into a bigfoot up there in the wilderness. Several soldiers were knocked to the ground with the creature standing over them before it ran away." Steve went on to tell of another interesting incident while on his second tour of duty in Vietnam (1971–72).

> Some G-75 Rangers (fourteen of them) were on commission in the mountain ranges above Que Sohn when the man at point (head of line) shot at a tall, furry, two-legged animal standing upright. Then a band of the strange animals came out of nowhere and attacked the fourteen soldiers. They had broken arms and limbs and looked like they had been beaten to a pulp. You can probably look this incident up in military archives, as some of these men were hurt so bad they had to leave military service. The animals were called, *Nguoi Rung* which in Vietnamese means, jungle man. (Source: witness Steve Bray.)

Four Hunters Killed, 1970s

An Oregon State Police officer in Deschutes County, Bend Oregon, was interviewed on his investigations. He commented on four hunters being killed in the Bend area in the 1970s. The bodies of the men were found busted and broken, and their rifles had been severely twisted out of shape by something with incredible strength. (Source: anonymous witness.)

Terrified Campers, 1970s

Summer time in the late 1970s, fifty miles west of Aberdeen, Washington, on the north shore of Lake Quinault in Grays Harbor County: A group of campers saw a sasquatch near their camp at July Creek. One of the men had a .22-calibre rifle. Frightened, he shot at the creature. The man didn't aim at or hit the sasquatch; he just wanted to scare it away, which he did. The campers were shocked at seeing the creature, but figured they were safe, as it was now long gone.

Later that night, four sasquatch showed up at the campsite and began throwing rocks and wood, and charging at the campers. The man with the gun fired at them, but only made matters worse. The sasquatch continued throwing things and then proceeded to beat the fenders off of their truck. The terrified campers built fires all around the campsite and burnt everything they could to keep the creatures away from them—all their firewood, as well as all their camping gear to keep the fires going.

They fought the creatures all night long. Just before daylight they made it to the truck and tried to get away, but the creatures kept them from leaving by picking up the truck and almost rolling it over.

When dawn broke, just enough to see, the creatures stopped the harassment and departed back into the woods. With what was left of the damaged truck, the campers escaped and headed straight to the ranger station where they called the sheriff's department to report the incident. (Source: International Bigfoot Society.)

Chapter 5 continues on page 321.

This image from frame 352 of the Patterson/Gimlin film, along with eleven other images from the same film, are the clearest images ever obtained of what is believed to be a bigfoot.

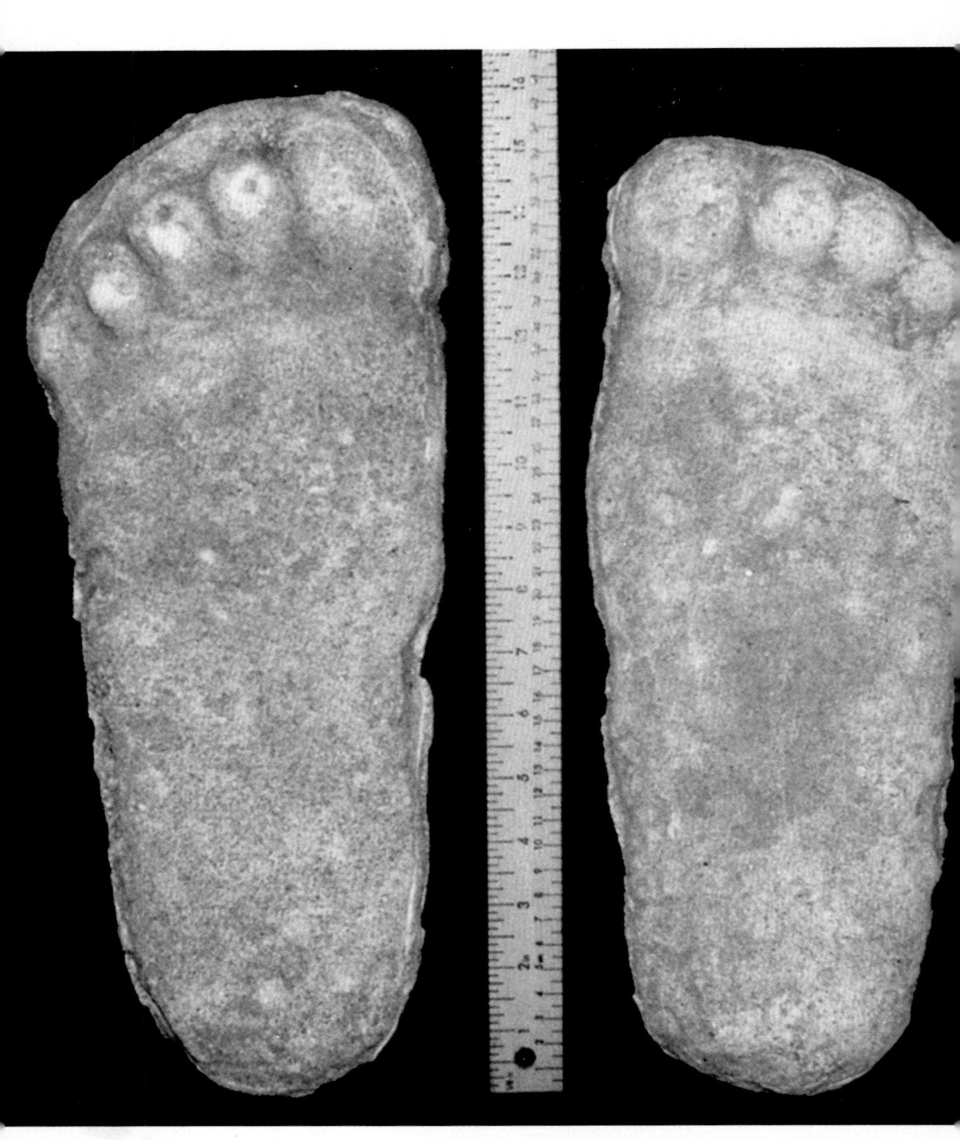

These plaster casts are direct copies of the casts Roger Patterson took at the Bluff Creek, California location where he and Bob Gimlin filmed a bigfoot on October 20, 1967. The footprints measured about 14.5 inches in length; the casts are slightly larger, which is commonly the case with castings. The casts and the creature filmed have been intently studied by a number of scientists, and some are of the opinion that the creature was a natural primate.

This is a scale model of the site at Bluff Creek, California where Patterson and Gimlin took their famous film. Patterson was standing in the position indicated by the red pin. He was about 102 feet away from the creature, which turned and glanced at him momentarily, and then continued at an angle into the forest seen in the top right-hand corner.

An example of what is believed to be a sasquatch footprint in soil. It is about 15.5 inches long. This print (along with others in a series) was found near Terrace, British Columbia in 1976. Bob Titmus made casts of a right and left foot, and is seen on the right displaying his work.

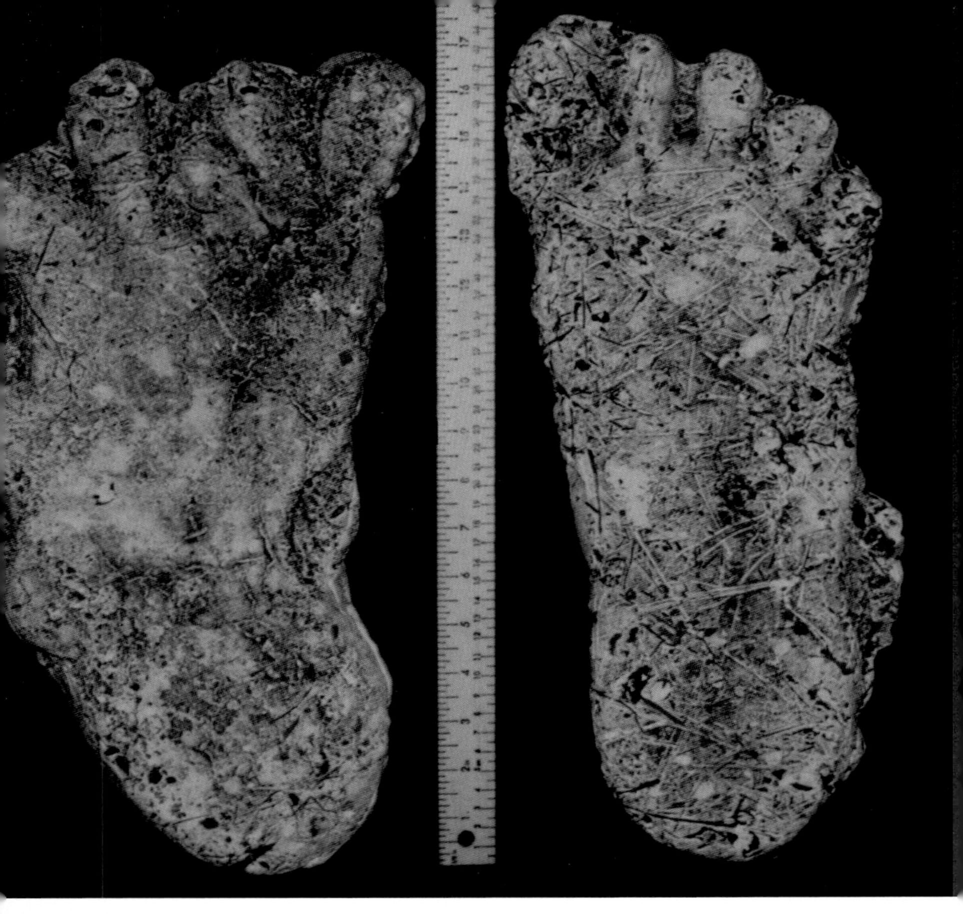

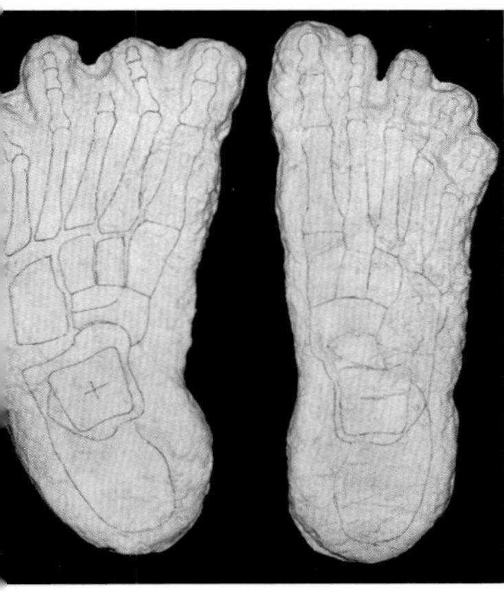

Plaster casts of footprints from a bigfoot creature who was believed to have had a crippled or deformed right foot. The casts were taken in Bossburg, Washington, in 1969 by René Dahinden. A trail of over 1,000 prints in snow was followed, and at one point it was observed that the creature apparently stopped and urinated. The casts have been studied by scientists, and the bone structure speculated, as seen on the right. The nature of the deformity shown was deemed too good to have been hoaxed. In other words, a hoaxer would have needed in-depth knowledge of anatomy to design a fake foot of this nature.

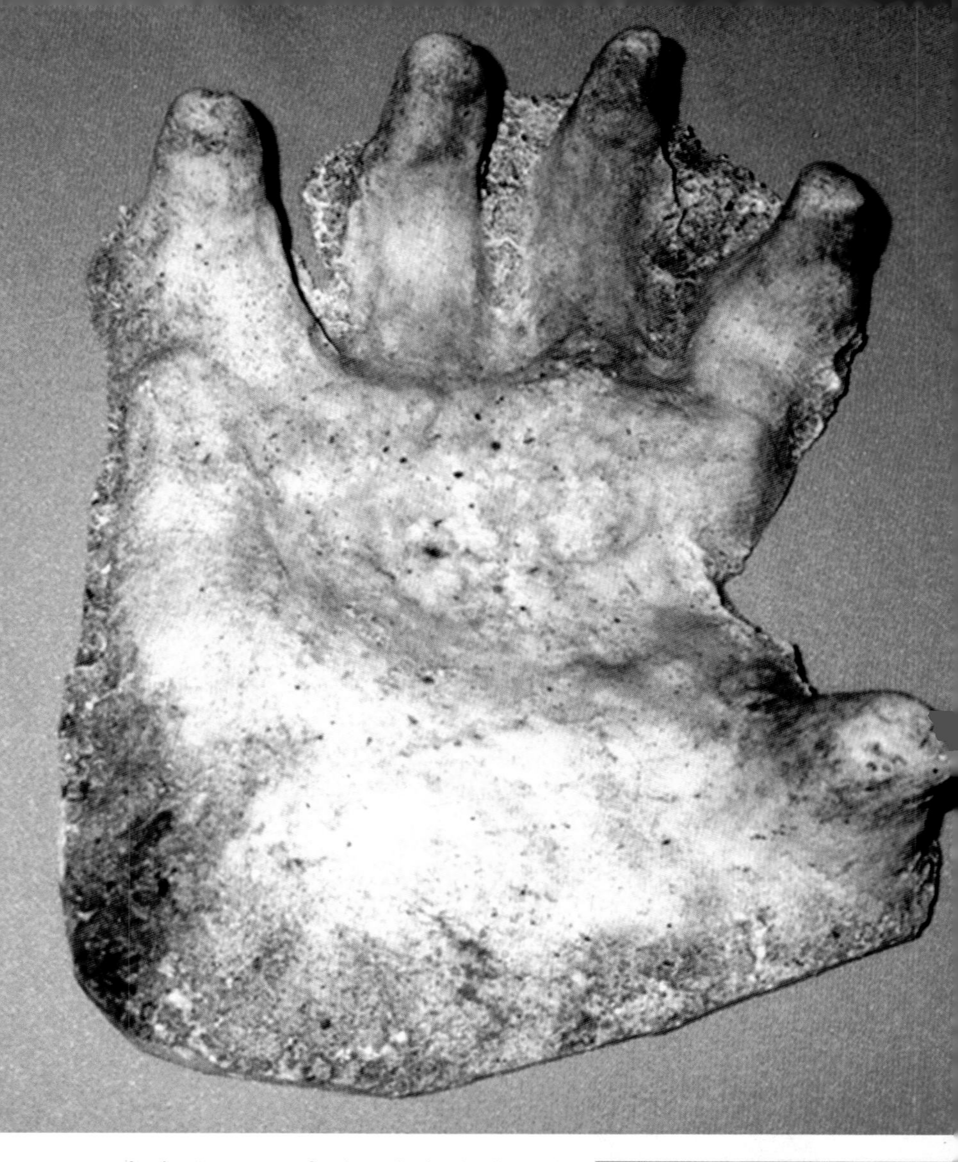

A plaster cast of a handprint believed by some professionals to have been that of a sasquatch. Its relative size can be judged by its comparison on the right with the hand of a large man, about six feet tall. The cast was taken by Paul Freeman in the Blue Mountains, Washington in 1995.

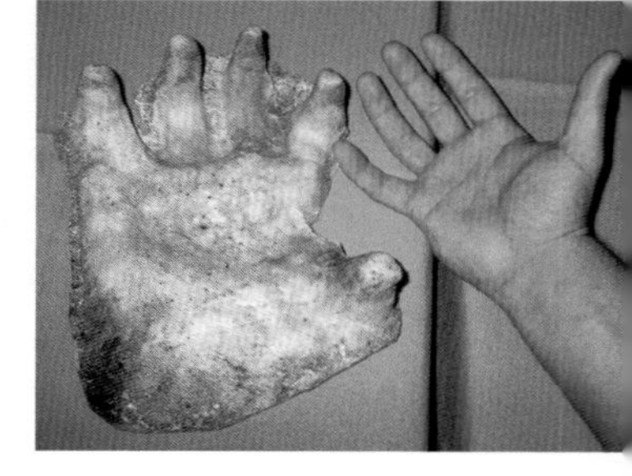

Hand cast taken by Bob Titmus on Onion Mountain, California, in 1982. The print was at the bottom of a shallow pond, and Titmus drained the pond to make the cast. The length of the cast from the end of the palm to the tip of the longest finger is twelve inches. The same measurement for a typical six-foot-tall, 200-pound man is eight inches.

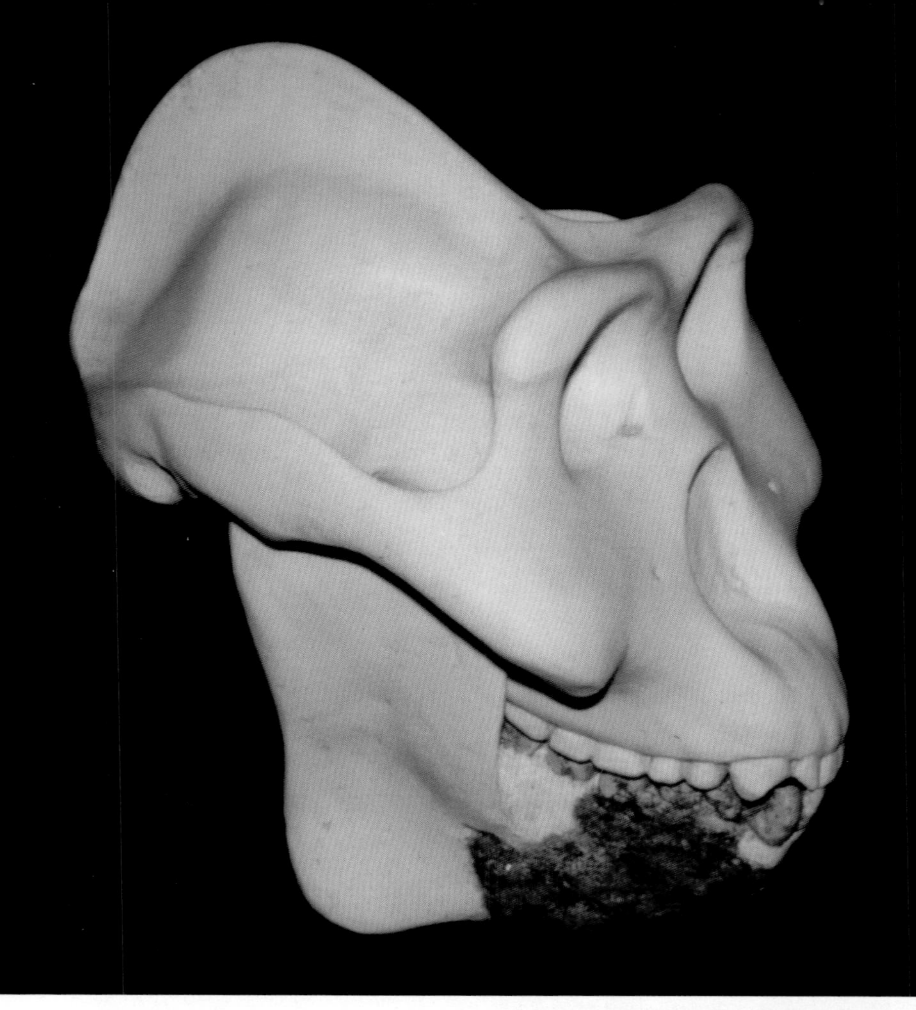

Speculated skull of a *Gigantopithecus* constructed by Dr. Grover Krantz, based on an actual jaw bone. The relative size can be seen on the right where it is compared to a gorilla and human skull. The "giganto," a large prehistoric primate that lived in Asia, is considered a "candidate" for the sasquatch. It is thought that the creature may have migrated to North America over the Bering Strait land bridge, and somehow escaped the fate of its Asian predecessors.

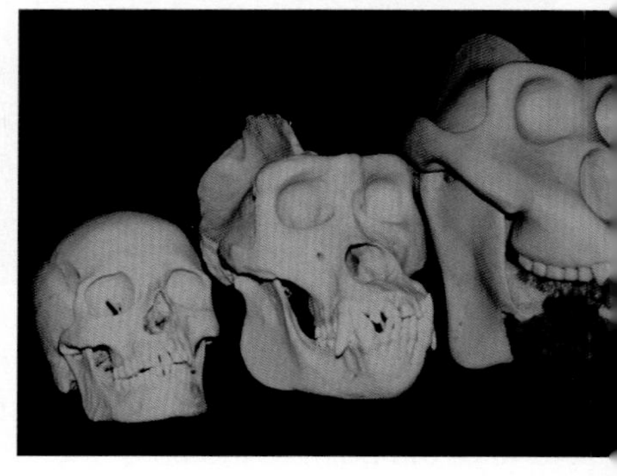

HUMAN GORILLA GIGAN

Possible sasquatch hair. At left a single strand at 260x magnification. Many hair samples have been submitted for analysis. One indication that the hair is not human is by the presence of "native terminations." In other words, neither end of the hair shows that it has been cut. Human hair always has one cut end. The only exception would be the hair from an infant who had not yet had his or her first haircut. The other indicator is the absence of a discernable medulla (inner or deep part). However, all we can conclude is that the hair is not human, or from any known creature. A confirmed sasquatch hair would be needed for positive identification, but such would, of course, make the comparison unnecessary if the object was to prove the creature's existence.

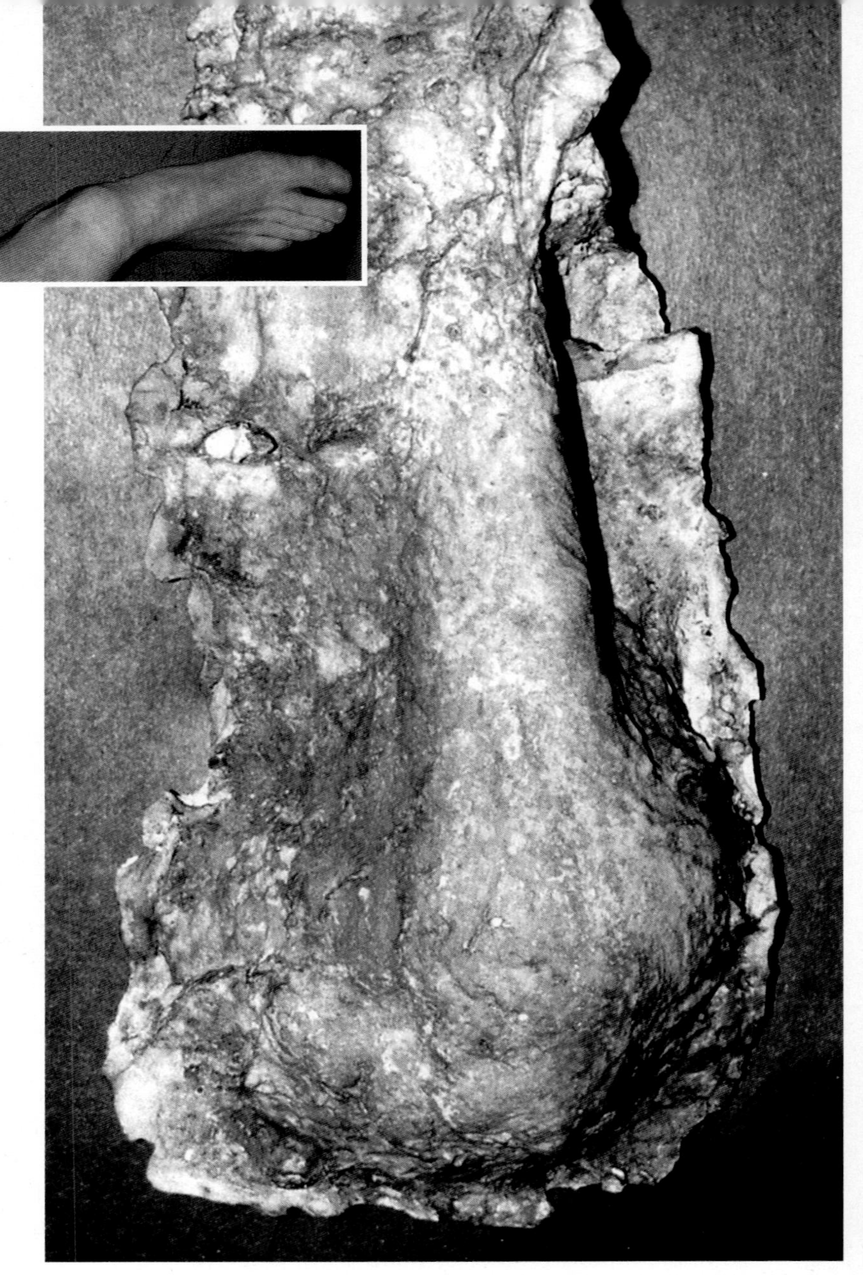

Cast of a heel impression and back of the leg, believed to have been made by a sasquatch. It is from the noted Skookum cast, which shows various body prints. The creature apparently dug in its foot as seen in the inset image. The heel is larger than a human heel and does not appear to be from any known animal.

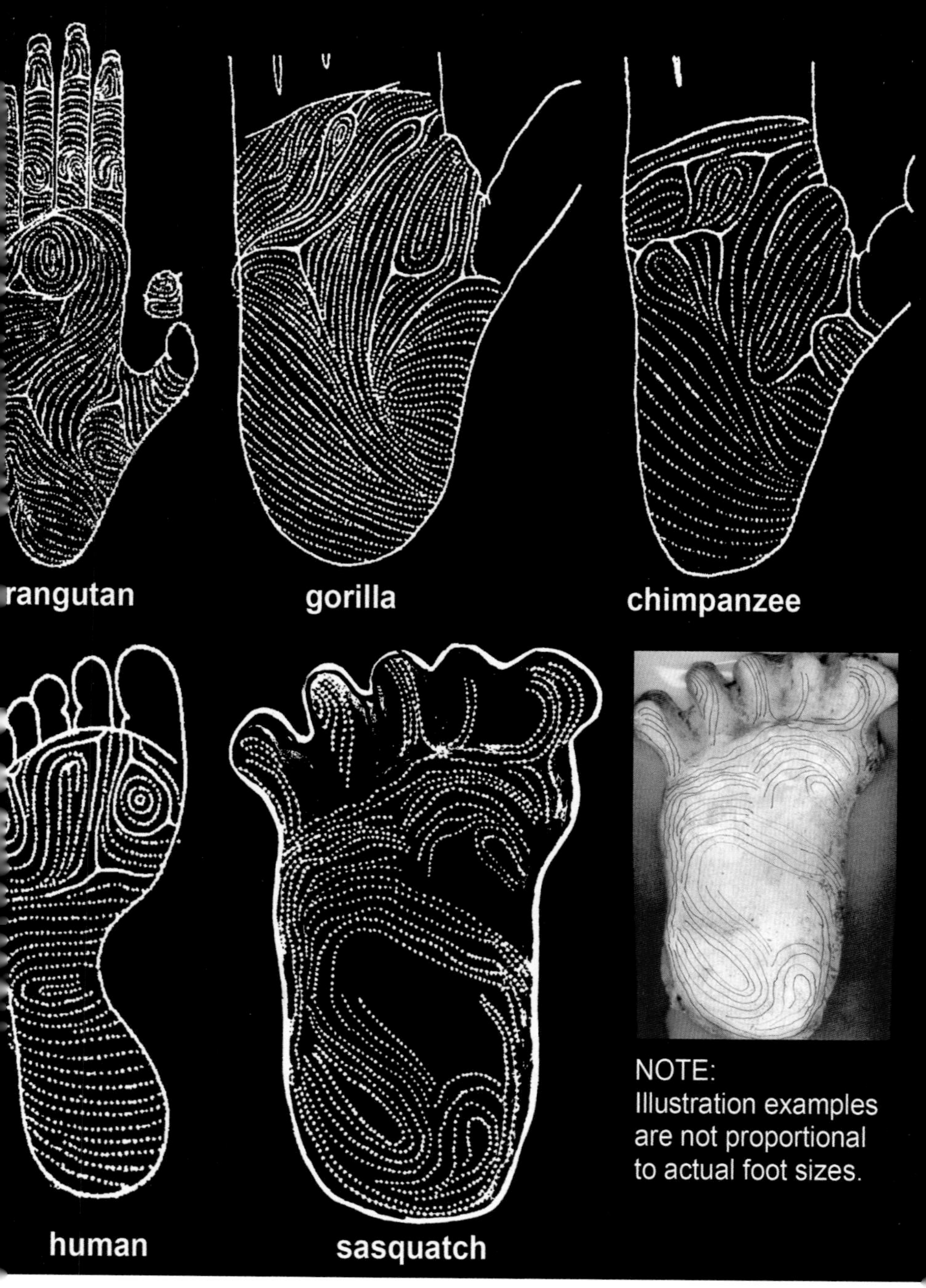

Dermal ridge pattern examples. The cast shown had the dermal ridges provided in the illustration for the sasquatch. It was taken by Paul Freeman in the Blue Mountains of Washington State in 1984.

This eyewitness drawing of a sasquatch has considerable distinction over drawings by regular artists. It was created by Carene Rupp (Northern California), who states that she has had numerous encounters with the creatures over many years. She has attempted on many occasions to get a photograph, but has only been marginally successful—images are too blurry to see any details clearly. Nevertheless, researchers, and at least one scientist, who have visited Carene are convinced that her encounters definitely occur, and she fully cooperates in helping to resolve the mystery.

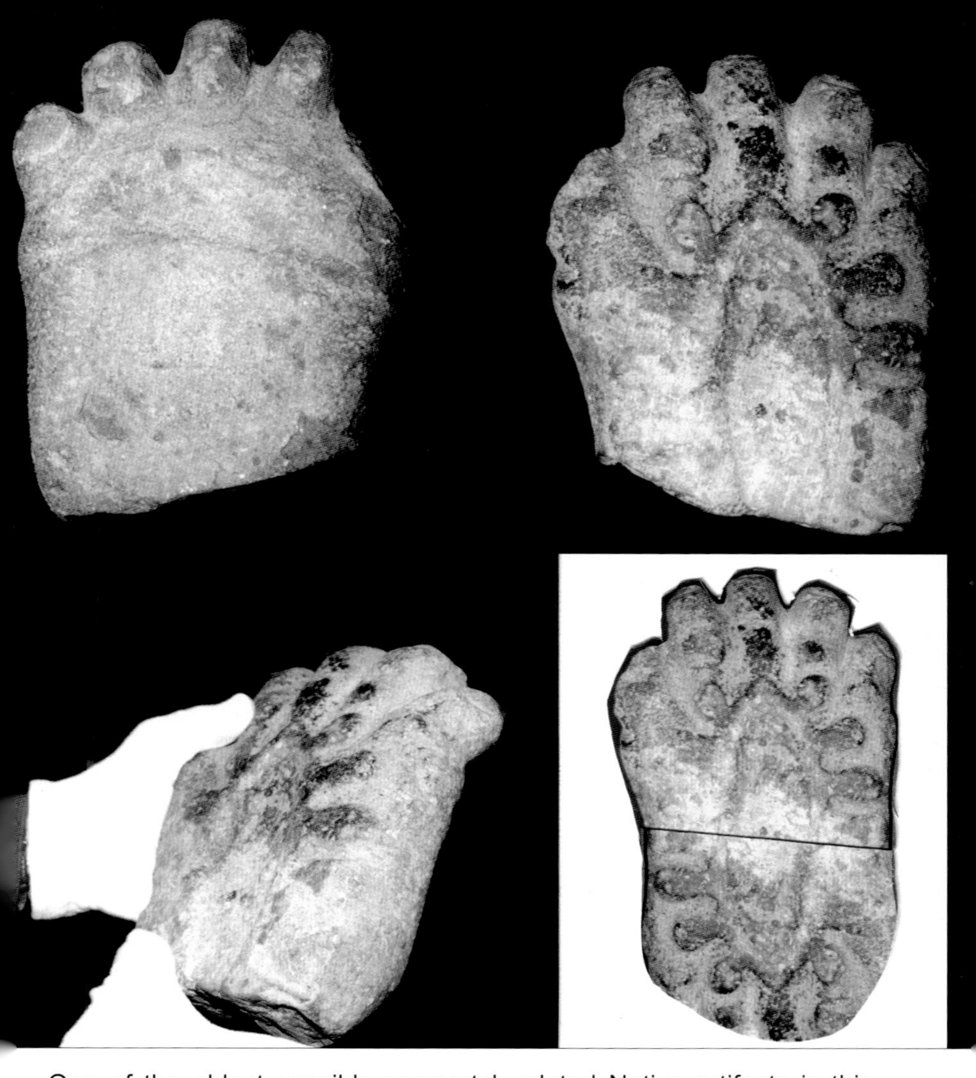

One of the oldest possible sasquatch-related Native artifacts is this stone foot. It was found near Lillooet, British Columbia in 1947. The big toe and lower portion are broken off. The last images shows a reconstruction that might indicate how the original foot would have looked (although still less the big toe). The artifact is thought to have been a medicine man's ceremonial stone. Perhaps the oval hollow was a bowl. As the sasquatch has sacred significance to many Native people, it is thought that perhaps modeling the foot after a sasquatch footprint was intended to give it some mystical significance. The actual age of the foot is not known, however, other Native stone carvings found are estimated to have been made between 1,500 BC and AD 500.

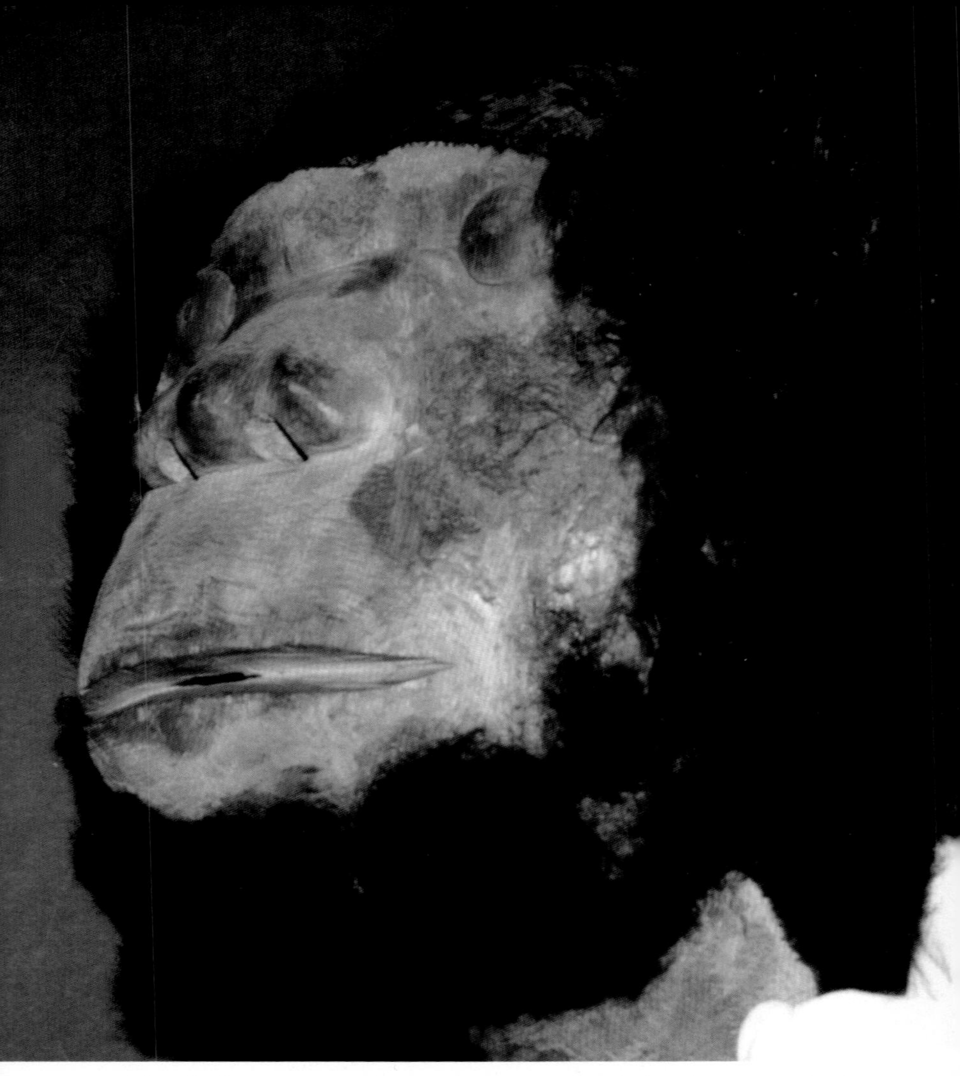

This lifelike sasquatch mask was carved by Ambrose Point, a Chehalis, British Columbia native, in the late 1930s. It is thought that Point saw a sasquatch and made the mask in the creature's image. When we compare the mask's features to other sasquatch sightings, there are remarkable similarities—large eyes, flat nose, wide space between the nose and upper lip, and thin lips. Point gave the mask to John W. Burns, a teacher at the Chehalis Reserve. Burns effectively coined the word "sasquatch," which is based on a similar native word, saskehavis, which means "wild men." Burns donated the mask to the Vancouver Museum in 1938, where it currently resides.

Although a highly prized artifact in its own rite, this Native mask is thought to be a copy of a significantly older mask. It was created by a Kwakiutl Native and represents the Buck'was, or "Wild Man of the Woods." The fact that it is not lifelike and is painted (rather than left natural, like the Chehalis mask) would probably indicate mythological aspects.

This Haida mask represents the Gagit, or “Land Otter Man.” It is also a copy of a much older mask. The spines in the creature’s lips are those of sea urchins and fish (dorsal spines) which the Gagit painfully endured or suffered when it ate such foods. This creature’s association with the sasquatch might be found in the fact that sasquatch are said to be extremely good swimmers, like the otter. It could be, therefore, that if native people saw a sasquatch swimming, they would associate it with an otter, and thereby call it an “otter man.” The word “land” would signify that the creature was also at home in the forest, far away from water.

Snooping Bigfoot, 1970s

In the early 1970s, a Ms. Grant lived next door to a police officer in Sweethome, Oregon. She said he came over one day, very shaken. He had responded to an early evening incident from Manns Ridge. People from California were moving into a new house on the ridge, and had heard noises coming from the garage. Going out to investigate, the family found a huge, tall, hairy creature stooped over looking at stuff in boxes.

It immediately left, stepped over a barbwire fence and loped off. The creature was described as foul smelling and of a really dark color. Silver and black hairs from the fence were analyzed by the Oregon State University and said to be of no known species. The police had been ordered to keep quiet to avoid a panic. The Californians never moved into the house. (Source: *The Track Record.*)

Steer Stolen From Barn, 1970s

In the 1970s, a report from Forks, Washington during an unusually heavy snowfall, told of a lady hearing a loud commotion in her barn. Investigating, she found the door open and a bigfoot who, while totally ignoring her, was carrying out a 300-pound steer. Apparently, researchers John Green and René Dahinden were notified and tracked the creature back into the hills. (Source: *The Track Record.*)

Rescue on Mt. St. Helens, 1970s

In the 1970s, two people had become stranded in a glacial crevasse on Mt. St. Helens. Four expert rescuers were sent to search for them. The rescuers were climbing the glacial ridge, and on the opposite glacial ridge, a Bigfoot was climbing parallel to them. It looked toward them, then descended into a vertical crevice in the glacier and was not seen again. The rescuers would have liked to have gone after the creature, but had to rescue the two-trapped climbers. All four of the rescuers saw the creature. (Source: *The Track Record.*)

Boxcar Bigfoot, 1973

In 1973, there was a big lumber company near Deep Lake, Washington (Stevens County) that made cedar shakes. An open railroad boxcar was on the siding, waiting to be loaded with shingles, when one of the lumber employees noticed a bigfoot inside. It was just looking around, with its back to the door. The employee snuck up, closed the door, and ran to get help. Of course, when he returned, the heavy plywood door was smashed to pieces, and the trapped critter was long gone. Huge footprints were found in the area. (Source: International Bigfoot Society.)

Night Raider, 1975

In the fall of 1975, a bigfoot was sighted frequently raiding the garbage dump on the Indian reservation at the mouth of the Nooksack River in Whatcom County, Washington. Furthermore, one rainy night, loud screams and loud pounding on picnic tables was heard coming from a nearby campground along the river. Not long after, that same night, the river flooded. In 1996, in that same area along the Nooksack River, a bigfoot had turned two mobile homes over. Nobody was hurt, and they were puzzled about the cause of the attack. This area has produced many bigfoot reports. (Source: International Bigfoot Society.)

Truck Driver Hits Bigfoot, 1977

At the end of January 1977, Doug McClure was crossing Satus Pass, Highway 97, ten miles northeast of Goldendale, Washington, near Three Creeks on the north slope. It was three in the morning and there was snow on the ground as, it had been snowing a little earlier. Doug was driving a tractor/trailer rig when he saw a bigfoot next to the road. When it moved onto the road, he was unable to stop, and hit it hard. It damaged the tractor, pushing the bumper all the way in to the front wheel. The bigfoot was not apparently badly hurt or frightened. It stood up, looked McClure in the eye through the front window of his truck cab, and then ran away on two legs.
(Source: International Bigfoot Society.)

Boy Scout Scare, 1978

Thad Byrd of Seattle recalls an incident in 1978 when he was eleven years old. He was in the Boy Scouts, and his group had gone to the Mines of Monte Cristo near Index, Washington (Highway 2, east of Monroe). He had wandered off while his companions were having a snowball fight; it was springtime, and during the thaw there was still some snow on the ground.

At one point, he looked up to see a big, hairy creature staring at him. It threw its arms in the air, and started a high-pitched screaming. Thad ran for the car, and locked the doors behind him. He said it had brown, shaggy hair, not thick like the Patterson creature (no hair around its face, and its arms were longer). It was seen from about fifty feet away. (Source: International Bigfoot Society.)

Bigfoot at the Ballpark, 1979

In the town of Yacolt Oregon (Clark County), the baseball field was the site of three bigfoot reports. In 1979, a reddish-brown bigfoot actually came from the nearby forest and stepped out onto the playing field briefly while a game was in progress. Everybody in the visitors' viewing stands saw it. The sun was shining on its hair; it was a clear sunny day, and the chestnut hair was reflecting richly.

Later, another sighting was made near the concession stand where several people saw it. The third incident, one man had apparently again seen it from across the clearing. The incidents were unexplained, except that bigfoot must be very curious. However, there was a rendering plant nearby that also butchered and cut meat, and this could have initially attracted the bigfoot. (Source: International Bigfoot Society.)

Man Sees Dead Bigfoot, 1980

The witness who provided the following report wishes to remain anonymous. He had read in Ray Crowe's *Track Record n*ewsletter about the cargo nets and helicopters used to transport dead bodies of bigfoot after the Mt. St. Helens eruption on May 18, 1980. He wanted to add his information in support this account. At the time, he was far east, in Spokane, Washington, visiting an aunt who lived in the "burbs" near Fairchild Strategic Command Air Force Base.

About 11:00 a.m., he says that a large, double-rotor, green helicopter, with a big star, went overhead about 100 to 150 feet, and he said he could see hairy arms and legs hanging out of the cargo net. The chopper was headed westbound (back toward Mt. St. Helens). He said he could see at least three creatures, along with tree debris and other garbage (apparently used to disguise the load). The creatures were still covered with gray ash.

"I got a real good look at them," he said. He immediately got on the phone and called several federal numbers, including the FBI, to get some information on what he had seen. Several days later, he had a return call from a deep-voiced caller, unidentified, except that he said he worked for the government. He told the witness, "Due to national security, if you ever tell anybody what you saw, unpleasant circumstances could develop. You might disappear, and nobody would ever know what happened to you..." He summed it up and ended the call with, "I'll erase your ass!" and, "There were no bigfoot or sasquatch parts in there!" (Source: *The Track Record.*)

More Bodies on Mt. St. Helens, 1980

Speaking of bigfoot bodies: credible rumor has it that the Army Corps of Engineers had taken out two bodies from Mt. St. Helens, two months after the blast (eruption). There was a dredging operation of the Cowlitz River and two bigfoot bodies were found in the sand. A chopper came and flew them off. The crane that was doing the dredging was from the Manatowaka Company. (Source: researcher T.R..)

They Like to Watch, 1980s

A woman named Mildred was involved in building a cabin twenty-five years ago near Moses Lake, Washington (Grant County). She said that five or six bigfoot creatures gathered on a hill opposite them and watched them working for hours, leaving her to believe that the creatures were peaceful. (Source: researcher D.D.)

Ranger Battalion Spots Bigfoot, 1984

An ex-military soldier, Ben, said that in February 1984 he was airlifted by a Blackhawk helicopter in a five-man team to a site four and a half kilometers south of Cat Lake, east of Sequim, Washington. The ranger battalion team was required to do a ground forest patrol (it was really cold—somewhere between midnight and 3:00 a.m). They proceeded in what Ben called a "lazy W" formation. As they moved forward, some "thing" made a weird primal scream—almost like a horse neigh. Apparently hearing the men, whatever it was that made the scream left rapidly. It was an old-growth forest—very dark—although the mens' eyes had become attuned to the darkness. They had Antas-Five night vision goggles, and two soldiers at the extreme ends of the "W" had PBS2-4 sniper scopes. They saw the creature run off on two legs. It was seven to eight feet tall and a good 500 to 600 pounds. (Source: International Bigfoot Society.)

Skunk Cabbage Snack, 1985

In October 1985, near Cumberland, Washington (King County) on the road to Palmer, two hunters trying to spot grouse ran into a strong smell, like that of a wet elk, coming from a swamp. Seeing something big move only 100 feet away and finally stand up from a squatting position, they realized it was a bigfoot. They rejected the idea of shooting the creature, opting instead to fire a gun and to scare it away. "It was eating skunk cabbage leaves and roots. It was very wary of us and kept rocking back and forth nervously." The creature would duck down every now and then, apparently to "hide." There was no noise except grunting like a big ape and a squeal when it took off as the gun was discharged.

Estimated at eight to nine feet tall and 400 pounds, it had grayish colored hair, twelve inches long all over its body. It had a blackish colored flat face with lots of hair, large lips, rounded head, and blackish skin visible on its face and hands. The eyes showed white corneas with black pupils. It used its arms to propel itself forward, grabbing at vine maples to swing behind as it retreated. They noted that the creature had thumb dexterity as it peeled back the leafy material on the skunk cabbage to expose the fleshy root and eat. The upper lip pulled back when it ate and it had a row of flat teeth in the front. (Source: researcher Thom Powell.)

Soldier Hears and Smells Creature, 1985

In late July 1985 at the Fort Lewis military base, south of Tacoma, Washington, a soldier was with his platoon being air-lifted by chopper for a week's maneuvering, somewhere close to Nisqually River and Mt. Rainier. He was assigned to a foxhole listening post in the forward area. It was around 2:00 a.m. when he heard heavy footfalls to his front to the right, heading away to the left. He then sensed a rotten cabbage or hot garbage odor, very putrid, making his eyes water. Forty-five minutes later, there was a scream from a couple of miles away that he couldn't describe, sounding painful, or even a bit lonely. The next morning the smell still lingered in the air. (Source: researcher Thom Powell.)

Peek-a-boo Bigfoot, 1986

Harold Morris and wife (non-believers in bigfoot) were collecting rocks at Opal Butte in a county park campground near Heppner, Oregon (Morrow County). They had parked their car and trailer at the campground. Upon hearing strange noises near their car, they looked over to see a creature hiding behind the front of the car. It would stick its brown, pointy, head up and look at them, then hide, over and over again, like "peek-a-boo." Finally, when tired of the game, it ran quickly up a bank and out of site. Harold's wife would not let him shoot it with his 30.06. (Source: *The Track Record.*)

Berry-picking Bigfoot, 1987

K.S. and four friends drove from Portland, Oregon, to the south side of Mt. Hood, to camp and hike a trail around the mountain. They found a nice spot at Sherwood Camp, near a creek, and surrounded by a good crop of huckleberry bushes. They set up their tents, all close together.

At 5:30 a.m. K.S. arose (the others still asleep) and noticed movement about seventy feet away in the berry bushes and evergreen trees. To her shock and surprise there was a large, thickset, seven-foot-tall, light-beige-colore,d female creature, entirely covered in hair, with its back to K.S. It was trying to reach something, a branch perhaps, that was about fifteen feet high.

About ten feet away was another smaller, hair-covered creature (no hair on the front of its hands, bottom of its feet, or around the eyes) and slightly darker in color—a dark beige. The hair on both creatures was about four inches long. The smaller creature was only about three feet tall and was bent over picking up a stick, which it was trying to put in its mouth. The mother gave a kind of grunt at the little one, apparently to tell it to leave the stick alone, and it dropped the stick.

K.S. moved slightly and stepped on a twig, thus making a "snap" sound. The mother creature turned to look right at her. She grunted again at the little one, reached down and took the young one's hand, motioning for him to come with her. She looked in the direction of K.S. again, grunted softly, and they were gone behind the trees. (Source: researcher Ray Crowe.)

Albino Bigfoot, 1988

In the summer of 1988, in Grays Harbor County, at Little North River, southwest of Aberdeen, Washington, a twelve-year-old boy was fishing for crayfish on a small creek. He had arrived here on his motorcycle. Feeling like he was being watched, he looked up to see a white bigfoot standing twenty feet away on the opposite bank. The bigfoot had a potbelly, bloodshot blue eyes, a pink complexion, a wide, flat, pink nose, and buckteeth. The hair was mostly white, (possibly an albino?), very white from the chin down to the chest and belly, grayer on the head, shoulders, arms and back, and gray from the knees down.

The creature was estimated to be over six feet tall. The boy threw the crayfish in the air and scrambled to his motorcycle. At the same time, the creature departed down the creek bed. It had a pronounced limp. The boy returned later with his father and they found huge footprints. The right footprints indicated that the creature was crippled. (Source: International Bigfoot Society.)

Family Attacked, 1989

A border patrolman and his family (wife and young girl) were camping near the Canadian border in Whatcom County (exact location not given), when they were attacked by one or more possible bigfoot. Both the patrolman and his wife were killed. One body was found wrapped around a tree, and another lay atop a large boulder. The campground was torn up, and the tent literally ripped to pieces. The little girl (three or four years old) was unharmed. She was found sitting at a picnic table. Sixteen-inch footprints were found all over the campsite. The Whatcom County Sheriff's Department was at a loss as to what to do with the case, so they called in the FBI, which, after investigating, kept all information confidential (sealed and hushed).

A game warden from the game department revealed that hired "predator hunters" are instructed to kill every bigfoot they see and to report killing to the department. The game department would then deny queries, because as far as bigfoot is concerned, it doesn't exist, and has never been classified, so there's no such thing.

When interviewed, the game warden said they had taken many photographs of tracks in the area, and when asked if the photos taken could be from a grizzly bear, he replied that these were certainly not grizzly tracks. (Source: researcher Ray Crowe.)

Rock-throwing Bigfoot, 1989

In March of 1989, Rick Jank was camping at Thunder Lake campground with another family in the North Cascades National Park, Washington. Two boys from the other family were fishing on Colonial Creek near Diablo Dam. Suddenly the boys returned, very upset, enough that they left all their fishing equipment behind. They said they had seen a bigfoot across the stream and it was throwing rocks at them. Their dad and Rick went to investigate.

Indeed there was a bigfoot that started throwing large rocks towards the two men from thirty-five to forty feet away. The rocks were large, sixteen to eighteen inches across. Rick and the family broke camp and left. Rick described the creature as over seven feet tall, narrow waisted, dark reddish-brown, three-inch hair, a high brow, and looked more like a monkey than a human. Rick said that the creature wasn't particularly aggressive, only splashing the rocks in the water as if to scare them away from his territory. (Source: International Bigfoot Society.)

Bigfoot in the Water, 1991

In 1991, two Indian boys, ages twelve and thirteen, were watching their salmon nets from a boat near the Second Tunnel on the Columbia River in Oregon. Both were sort of half asleep. One of the boys woke up sensing that something was wrong. He looked around and saw a bigfoot in the water, up to his chest, looking at him. The boy woke his brother and they both had a good look at the creature, which then reached forward, rocked the boat a little, and then just turned around and walked away. (Source: International Bigfoot Society.)

Easter Bigfoot, 1991

It was close to Easter in April of 1991. A family was living on a ranch out near Molalla, Oregon, where they raised cattle, chickens, turkeys, and pigs. Coming home one night, the headlights of their car picked up a rare, possible albino bigfoot as they turned into the long driveway. He squatted down behind some bushes trying to hide, but they had already seen him.

It was seven feet tall, with, eight- to nine-inch light-cream-colored hair. The mother commented, "Looks like the Easter Bunny's back again." This was the third year in a row that they had seen the white bigfoot, always around Easter time, and he would hang around for four or five nights in a row, judging from the dog howling. (Source: *The Track Record.*)

Dead Baby Bigfoot, 1992

In Molalla, Oregon, near Estacada, a hunter had found a dead baby bigfoot, a furry little brown female, that was left ten feet up in the boughs of a tree. It was then covered with other boughs—perhaps an "Indian style burial." The hunter's attention was called to the tree by the piles of huge droppings lying around the base. One can almost visualize a grieving female bigfoot staying near the remains of her dead young.

The hunter contacted the Portland State University, thinking they would be very interested in a new species of animal. He, of course, got the old horse laugh. They wouldn't even look. So the baby creature went into the hunter's deep freezer. The college later called back and asked to see the corpse, but the harm had been done, and he told them where to go. The hunter ultimately disappeared, along with the corpse. (Source: International Bigfoot Society.)

Family of Bigfoot, 1992

A Warm Springs Indian told of a bigfoot story related to him by his Yakima Indian grandfather. It seems every fall, a family of six bigfoot passed near his place on Rattlesnake Creek, in Klickitat County, Washington. They were heading west from a hilly, forested area east of him. He knew it was the same family, because one of the female creatures had tan eyebrows that stood out.

In about 1992, a man by the name of Garret went to the same area to see if he could see the creatures, and was rewarded by seeing one at two o'clock in the morning when barking dogs awoke him. They could be heard barking from far away, but became silent as the creature approached. He saw the creature in silhouette standing behind an outhouse before it faded into the darkness. It was very quiet, but a stinking stench was obvious. (Source: Garret S.)

Bigfoot Funeral, 1992

The eyewitness of this story wishes to remain completely and securely anonymous, and so shall be called John.

John, a man well educated and a teacher of philosophy, had an extremely rare and strange sighting in the summer of 1992, east of Estacada, Oregon. He had gone on a long hike by himself, finding it very peaceful, enjoyable, and uneventful. About four or five hours into the hike, the silence of the forest was interrupted by the sounds of "clink, clink, clink."

Curiously and quietly approaching the upstream noise, John was shocked to find himself less than a hundred yards away from two bigfoot creatures. Startled, but very intrigued, he hid behind some bushes to watch.

One was larger than the other, but he couldn't tell if they were male and female—only that they were tall, huge, and covered in dark hair. As he continued to watch in awe, he could see that they were piling good-sized rocks on top of something, thus accounting for the "clinking" sounds.

Taking a closer look, he realized what they were piling the rocks upon. The bigfoot were engaged in burying another dead bigfoot under a pile of stones. They had not dug a hole, but were simply covering the body with stones. Suddenly, two smaller, red-colored bigfoot appeared, each helping to place a stone. John said that the bigfoot were acting very "sad," and for some reason he himself experienced intense feelings of sorrow as well. He watched them for as long as he could, before quietly slipping away. (Source: anonymous witness.)

Hot Foot, 1992

In the summer of 1992, David was fighting fires with the Job Corps on the Yakima Indian Reservation in Washington State. During a fire, he witnessed a bigfoot that came running out of the burning area, running right through the hot coals and toward several Indians on the fire line. Three of them dropped their equipment and left the area, refusing to go back. (Source: International Bigfoot Society.)

Aging Sasquatch, 1993

Dan B. and a party of eight were fishing for steelhead in the spring of 1993 on the Clallam River, about two to three miles from the Strait of Juan De Fuca, west of Port Angeles, Washington. After dark, as they were eating dinner, they heard movement in the brush, then a crunching on the gravel road. Dan turned the pickup headlights on, and there was an old, gray-haired bigfoot in the beams, mouth open in surprise. The teeth were broken and decaying, and it had a beard. The chest hair was ragged looking—black, though mostly gray. In a panic, it ran through a vine maple stand, breaking limbs, and cut a swath across a field, screaming as it ran. Human-like tracks were later found that measured sixteen inches long. (Source: researcher Fred Bradshaw.)

Goat Found Dead, Bigfoot Suspected, 1994

February 1994, Welches, Oregon (Clackamas County): A farmer reported that a twenty-five- to thirty-foot section of his barbwire fence was down and he was missing a goat. The goat was later found dead, its neck broken, bowels missing, and tongue gone. The next day, the body had been moved and was hanging over the low limb of a tree. On the ground surrounding the tree were sixteen-inch, five-toed tracks. A researcher showed up to investigate and made several plaster casts of the prints.

This account makes one wonder about animal mutilations with speculated UFO connections. Is is possible bigfoot might be responsible for some of them? .(Source: Fred Bradshaw.)

Monthly Visitor, 1994

In 1994, West of highway 101, on Bunker Road, Grays Harbor, Washington, a resident reported that a bigfoot came around once a month. Large tracks were found, though the creature was never seen. It went to a burn barrel where the resident noted meat wrappers were thrown around and the potato peelings would be missing. The resident began leaving whole potatoes and cucumbers out, and soon found out that cucumbers were its favorite. He put up a night-light, but the creature would only come around when it was off. (Source: researcher Fred Bradshaw.)

Wynoochee Lake Wrestlers, 1994

Most interesting is an incident that took place in the spring of 1994, at Wynoochee Lake in Washington State. For a good thirty minutes, a group of campers from different parties watched intently through a spotting scope as two bigfoot wrestled with each other on one of the nearby mountainsides. (Source: researcher A.P.)

Yakima Protect Bigfoot, 1995

A Yakima Native American and traditional storyteller told of a friend driving from Portland to Warm Springs, Oregon, and stopping to rest. A bigfoot stepped out of the woods, came up to his car and began to shake it. It then turned away and walked back into the woods. Around the same time, on the Yakima Reservation, a bigfoot came down from the mountains and looked into trailers and houses. The Yakima Indians don't tell anyone their stories, as they're afraid bigfoot will be shot. (Source: *The Track Record.*)

Bigfoot in the Mineshaft, 1996

A man and wife were exploring an old mine shaft near the upper Clackamas River in January 1996. There are about thirty abandoned quicksilver mines in the vicinity (south side of the river). The mine being explored by the couple is near Lake Harriet, off Highway 57, east of the Ripplebrook ranger station. The horizontal shaft was boarded up with a special door for entry that one had to crawl through.

Using big flashlights, they entered the mine, the lady first. A short distance from the entrance of the 400-foot shaft she saw a bigfoot. She was so startled that she dashed back out with her husband (who did not see it) following. She then told him of the shaggy, silvery gray creature she had seen.

The creature was sitting hunkered up, head between its knees, and its arms wrapped around itself, with one human-like hand shading its eyes from the flashlight. She described it in human terms, and says she commented on the foul odor when they entered the shaft. Later, her husband returned with some neighbors, They, re-entered the shaft and explored it to its end without seeing the creature. They did, however, find a "bed" that was comprised of sticks, moss, and ferns. (Source: International Bigfoot Society.)

Sasquatch Rescues Man, 1996

Hav Tranh was hiking in the Deschutes National Forest Oregon, on July 13, 1996. He had been hiking up a steep slope, when he slipped and fell, breaking his leg, a compound fracture with the bone sticking out. He passed out, and when he regained consciousness, there were two bigfoot hovering over him. The larger one was seven to eight feet tall, a gray-white color. The eyes were black. It had a large, sloping forehead with a domed shaped head ending in a peak. And it had very big feet.

The second creature was the same height, but was a sandy-gray color. It had a white ruff on top of its head. They were very hairy and he couldn't tell the sex of either. They were jabbering at each other, not using words, but indistinct noises. He passed out again, and did not reawaken until his wife, Gioking, shook him. She told him two ape-men had carried him out of the forest and deposited him near her. She said they were horribly ugly and added that they had very long hair except on the face, heads, and feet. She took him to the hospital for medical attention. (Source: *The Track Record.*)

Close-up Sighting, 1996

Kurt Armbruster, from Scotland, was visiting friends in Curlew, Washington, near the Canadian border. On March 16, 1996, he took a hike up nearby Mt. Vulcan, and close to the top, he found himself seventy-five yards away from what he thought was a bear digging in a ten-foot snow bank. It was 9:30 a.m. and the sun was shining on the animal, so, being a nature lover he wanted a closer look. At twenty-five yards the creature stood up, a full nine feet tall, and turned to face him, a large digging stick in its hand.

They stared at each other for a few seconds and then it walked off into a thicket and was gone. Not believing what he had just seen (and thankful that he was still alive), Kurt walked over to where the creature had been, and noted that it had been digging up a bush that had red berries on it, and there was a small stack of the berries on the ground.

Kurt emphasized that the creature was a male—it had a very long penis and testicles, like a small horse, he said. The creature had reached its left hand up to its head, and was missing an index finger. The hair all over its body was a short, brown-black color, matted at the shoulders and thighs, and softer in the chest area. It had no facial hair, or hair on its palms or the soles of its feet.

He measured the tracks it left using his forearm. They were twenty-one inches long and about seven inches at the ball. Its stride was about seven to eight feet. It curiously slipped into the brush without breaking any branches or making any noise. There was an awful stench. When it first turned and saw him, Kurt said, it looked surprised, and its almond-shaped eyes widened.

He noted that the eyes appeared blue with the eyeball white. It grunted at him, and he saw the teeth were a bright white, and like human teeth. The arms seemed to hang proportionately, longer than humans. It did not have a neck. The naked skin of the face was like a Caucasian, and the head came to a rounded point; it had a large, human-like nose, and heavy eyebrows. The mouth seemed ridiculously small. (Source: *The Track Record.*)

Bigfoot on the Beach, 1997

On October 11, 1997, a woman from a search and rescue group was hiking near Claylock, on the Washington Coast. She was on a twenty-eight-mile wilderness trail when she stopped and dropped away from her group to take a nature break. As she peered out of the forest at the ocean, she noticed about 200 yards away, a tall, brown, hairy bigfoot sitting on a log, looking out to sea, and casually dragging its hand through the sand, leaning down to do it.
After several minutes of watching, the creature suddenly stiffened, looked her way, and made eye contact. Bent over and with giant strides, it bounded off into the forest, making a lot of noise crashing through the woods. She didn't follow, or look for tracks, but hurried to catch up with her group to tell them what she saw. They didn't believe her. (Source: researcher Fred Bradshaw.)

Prospector Spots Bigfoot, 1997

In 1997, an older fellow was panning for gold several miles up on Engals Creek in Chelan County, Washington. The valley where the creek runs is called Nightmare Valley. As he was focusing on his task at hand, he felt what he described as a "presence." Something was watching him. As he looked up, about twenty feet away and across the creek, stood a tall, hairy-looking, hominid male. The creature did not startle him—just more or less amazed him. The two of them made eye contact for what seemed to him to be about thirty seconds, and then the creature appeared to lose interest and just turned around and disappeared into the brush.

He described the creature as having a flat face, long arms, small waist and a large chest. The fellow decided he had done enough for that day, so he packed up camp and left before nightfall. (Source: *The Track Record.)*

Spotted from the Air, 1999

T.S. is a skydiver, and in August he was in Kapowsin, Washington, at the annual Skydive Boogie. When he and his friends were en route to the "drop zone" in their Cessna, they looked down and saw a sasquatch walking along a road. The pilot whipped the plane around and buzzed the creature at about 100 feet. It took off running down the road with huge strides. It later leapt off the road and ran up a hill—jumping stumps as if they were nothing. They came back for another look, but it was gone. (Source: researcher Thom Powell.)

Bigfoot Hit and Killed, 2000

In 2000, a fellow who worked for the County Roads Department told of a bigfoot being hit and killed by a car near Yale, Washington in Cowlitz County. He had to assist at the scene by blocking the road while a forest service chopper picked up the body. (Source: International Bigfoot Society.)

Mysterious Rock Pile, 2001

I have a co-worker who went camping with some friends near Mt. Hood in Oregon. They hiked a number of miles off the main roads, and came to an area with six piles of rocks in various locations throughout the area They set up their camp in this area and later knocked over the rock piles, thinking they were victims of a prank by some past campers who were used the same site. The following morning they awoke to find all six-rock piles had been restored, and a seventh rock pile was now present, next to their burnt-out campfire.

The ground was hard, and although they found impressions in the earth, they could not make out any definite tracks. (Source: researcher Dennis Fenton.)

Note by Ray Crowe: There is a report of a man in Klamath County, Oregon who was making a road. He came to a large ring of boulders, which he shoved aside with his cat. The next day, they were back in position. This happened several times before he gave in and built the road around the replaced boulders.

Shotgun Sally, 2005

Anonymous Report: Sightings continue to occur out near Vernonia, Oregon. There is a woman known as Shotgun Sally who lives in a trailer camp near a fishing pond off the Vernonia Highway near town. She regularly has contact with these creatures, as do the locals in her community. They apparently stink and make particular noises. Shotgun Sally has a tendency to shoot her gun off at night to scare these creatures away from her isolated home. They don't hurt anyone according to one of the locals who owns the fishing hole. (Source: anonymous witness.)

Notable Quotes: Sean Fries, 2007

> **Scariest Moment:** My scariest experience, hands down, was last summer (2006) in Six Rivers National Forest, at Onion Lake. I was walking around the lake checking the shoreline for tracks (the whole lake is about seventy-five yards across). I was on the opposite side of the lake from my truck, unarmed, and looking down at the mud, when I got that very strong feeling that someone was standing behind me. All the hair on my arms and back of my neck stood up on end and every part of me was saying, "Do not look behind you, just walk away." And that's exactly what I did. When I got back to my truck I then looked back and it *[whatever had been behind me]* was gone."

Bigfoot Has a Sense of Humor: We have an area of active study where we know there is a lot of bigfoot activity. We set up a bait pile in the middle of a big, muddy marsh on this old logging road. Nearby there was a pine tree and we figured it would be a great spot to install a camera trap. We set up the whole thing, put out the apples, set up the camera, left and came back a week later. Walking up to the spot, we were very excited as all of the apples were gone, but there were no footprints in the mud. We looked over to the camera, and it was turned completely around pointing away from the bait pile, with not one picture taken! These things are really smart and I believe very playful as well.

Tree Knocking: On June 9th, 2007, my fiancee and I went to the The Six Rivers National Forest, California. We were going out there later to camp for a few weeks and we went up early to put out some camera traps. When we got to Onion Lake, I found two thirteen-and-a-half-inch tracks along the shoreline. As I was casting the tracks, some tree knocking sounds began on the slope to the north of us, and very close. I started tree knocking back and whatever was knocking would match the number of knocks I made.

When I knocked four times, I got four return knocks; five times, I got five back. Then I made two knocks, and this time the returned knocks came from the southwest, so there were at least two of "them." I tried scanning the trees with my binoculars and they stopped knocking.

My fiancee, Kandee, started walking up the hill to where the first knocks came from, when there was a loud limb "snap," and she heard something moving away from her in the brush. We left to go set the camera traps and then went back to the shoreline in about an hour to retrieve the casts. During this time the tree knocking began again.

Chapter 6

Things to Know

> If you are out in the woods and come upon a bigfoot, do not be afraid. Look the creature directly in the eye, and like a spark of lightening, a jolt of energy will jump between you and the beast; and you will acquire all the secrets of nature.
>
> —Karok Indian tribe of Northern Califomia

Signs

Based on eyewitness accounts, the following is a list of things to look for in the woods that might indicate the presence or passage of a bigfoot.

- The woods are noticeably quiet; no sound of birds or other wildlife creatures. Eerily quiet.
- Large, human-like footprints.
- The overwhelming feeling of being watched;0oOr a feeling of dread or intense fear.
- If you bring along your dog, he shows signs of fear, refuses to travel any further, runs to hide, whimpers, or acts totally out of character.
- Something unseen seems to be pacing your movements.
- Loud crashing sounds through the woods.
- Trees or limbs with signs of being broken at seven feet up or higher.
- Rocks apparently being tossed and falling into your area.
- The sound of rocks being rapped together, or the sound of wood knocking (trees or logs being rapped with a large branch).
- Stacked rocks/boulders on or alongside a trail; rocks on top of tree stumps.
- Fallen trees blocking a path or trail.
- A bed of forest material (grass, twigs, leaves) that appears to be a resting place for a large animal.

- A large pile of scat (dropping) that appears human-like but is both too much for a human and too large for a human body.
- Loud shrieks, howls, whoops, generally in the dark, that you do not recognize as coming from a known animal. Also, unusual whistling.
- An unusual and unpleasant odor/smell, sometimes described as that of a wet dog that has rolled in something (fully discussed in the next section.)

Odor

A strong and sometimes overwhelming unpleasant odor is occasionally associated with a bigfoot encounter or its possible presence. What is the odor? Why do they smell so badly? Most of the following information is edited from a paper written by Dr. Henner Fahrenbach. He notes an observation of mountain gorillas as recorded by anthropologist Dian Fossey. She describes male gorillas as producing an "overpowering, gagging fear odor" either when fleeing from enemies, or when in confrontational encounters with other male gorillas. At eighty feet the smell is very intense, and may be coincidentally accompanied by a discharge of diarrheic stool.

> The origin of the odor seems to be in the axillary gland, a mass of apocrine sweat glands many layers deep in the armpit. Noted in the autopsy of a male gorilla, one gland was reported to be smelling and the other not, an indication of neural control over the discharge. (The gorilla had been shot, presumably on one side).

As humans, we may find ourselves exposed to intense odors such as those of a horse or cow without much of a reaction. However, we tend to be very offended by primate aromas, such as the smell of the ape house at the zoo or the smell of a sasquatch.

There are repeated reports of people experiencing an overwhelming feeling of being watched, being overcome by fear and abruptly leaving a campsite, forest, or other area to head for their cars or homes. This immediate emotional response may conceivably be due to a hormonal component of sasquatch sweat that is being perceived at an unconscious level.

Names

The following are some of the names by which bigfoot, or bigfoot-like creatures, are known throughout the world..

North America (U.S. and Canada) Common Names

Name	State/Province
bigfoot	All areas in the U.S.
Fouke monster	Fouke, Arkansas
grassman	Ohio
momo	NE Missouri
mountain devils	Pacific Northwest
old yellow top	Ontario (see note 1)
sasquatch	All areas–Canada (see note 2)
skunk ape	Florida
windigo	Quebec
woods devil	New Hampshire
wookie	Louisiana

Note 1: The reference here is to a particular sasquatch with a blond patch of hair on its head and a light-colored mane. It appeared and reappeared around Cobalt, Ontario over a period of sixty-four years. The last sighting was in August 1970.

Note 2: This is an anglicized Indian name meaning "hairy giant."

Northwest North America Native Names

Name	Meaning	People
At'at'ahila	not known	Chinookan
Bogs	not known	Bella Coola
Buck'was	wild man of the woods	Kwakiut
C'amek'wes	not known	Lummi
Choanito	night people	Wenatchee
El-Ish-kas	not known	Makah
Hecaitomixw	dangerous being	Quinault
Itssuruqai	cannibal monster	Shasta
Kala'litabiqw	not known	Skagit Valley
Madukarahat	giant	Karok
Omah	not known	Yurok
Qui-yihahs	the five brothers	Yakima/Klickitat
Sasahevas	wild man	Chehalis
Sasahevas	wild man of the woods	Salish
Sc'wen'ey'ti	tall, hairy, smells like burnt hair	Spokane
See'atco	one who runs and hides	Coast Salish
Sesquac	wild man of the woods	Coast Salish
Skanicum	stick indians	Colville
Skukum	devil of the forest	Quinault
Skookum	evil god of the woods	Chinook
Sne Nah	owl woman	Okanogan
Steta'l	spirit spear	Puyallup and Nisqually
Ste ye mah	spirit hidden under cover of woods	Yakima
Tsialko	wild Indians	Puyallup and Nisqually
Yayaya-ash	the frightener	Klamath

Note: Entries shown as "not known" simply indicates that I do not know the English meaning. There would definitely be a translation, but it was not shown in the material I studied.

Non North American Countries

Name	Country/Region
Almas	Mongolia & Russia
Biabin-guli	Europe
Barmanou	Afghanistan and Pakistan
Barmanus	Northwest Pakistan
Brenin llwyd	Europe
Chemosit	Africa
Chuchunaa	Siberia
Ebu Gogo	Flores Islands of Indonesia
Firla Mohir	Europe
Gin-sung	Asia
Grendel	Europe
Higabon	Japan
Kakundakari	Africa
Kapre	Philippines
Kaptar	Russia
Kikomba	Africa
Maricoxi	South America
Mapinguar	Brazil & Bolivia
Mecheny	Asia
Mirygdy	Asia
Ngoloko	Africa
Nguoi Rung	Vietnam
Nyalmo	Himalayas
Orang Pendek	Sumatra, Indonesia (Note: This is a small man-like creature.)
Orang Mawas	Malaysia
Tano giant	Africa
Teh-lma	Tibet
Ucumar	Latin America
Woodwose	Medieval Europe
Yeren	China
Yeti (or Meh-teh)	Tibet
Yowie	Australia

The Seeahtik Belief

The following newspaper article, which appeared right after the Mt. St. Helens incident involving Fred Beck and his party of prospectors in 1924, sums up the unique belief of the Clallam Native people on the nature of sasquatch/bigfoot. Although it brings into play various aspects of the paranormal (usually seen in Native lore), it is nonetheless very compelling. It is not beyond reason that the sasquatch could be a race of aboriginal people as the Clallam Natives claim.

The Seattle Daily Times
July 16, 1924

Clue to "Gorilla Men" Found. May be Lost Race of Giants

Clallam Indians tell of Eight-Foot Seeahtiks Who Killed Game by Hypnotism. Existence Kept Secret by Other Tribes

1924—Hoquiam, WA—Wednesday, July 16. "Mountain Devils discovered at Mt. St. Helens near Kelso, are none other than the Seeahtik Tribe," said Jorg Totsgi, Clallam Tribe Editor of *The Real American,* an Indian national weekly publication in an interview here today. "Seeahtik" is a Clallam pronunciation. All other tribes pronounce it: "Seeahtkoh."

"The Indians of the Northwest have kept the existence of the Seeahtiks a secret. Partly because they know no white man would believe them and the Indian, known for his honesty and truthfulness, does not like to be called a liar, and partly because the North-Western Indian is ashamed of the Seeahtik Tribe," said Totsgi.

The "Mountain devils" or "gorillas" who bombarded the prospector's shack on Mt. St. Helens, according to the description of the miners, are none other than the Seeahtik Tribe with whom every Indian in the North-West is familiar," said Totsgi.

Were Thought to be Extinct

The Seeahtiks were last heard of by the Clallam Indians about fifteen years ago, and it was believed by the present day Indians, that they had become extinct. The Seeahtik Tribe make their home in the heart of the wilderness on Vancouver Island and also on the Olympic Range.

As described by the Clallam Indians, the Seeahtiks are seven to eight feet tall. They have hairy bodies like the bear. They are great hypnotists, and kill their game by hypnotism. They also have a gift of ventriloquism, throwing their voices at great distances and can imitate any bird in the Northwest. They have a very keen sense of humor," Totsgi added.

In the past generations they stole many Indian women and Indian babies. They lived entirely in the mountains, coming down to the shores only when they wanted a change of diet. The Quinaults claim they generally came once a year to the Quinault River, about fall. The Clallams say they favored the river area near Brinnon on Hood Canal. After having their fill of fresh salmon, they stole dried salmon from the Indian women.

The Seeahtik Tribe is harmless if left alone. The Clallam Tribe, however, at one time several generations ago, killed a young man of the Seeahtik Tribe to their everlasting sorrow, for they killed off a whole branch of the Clallam Tribe but one, and he was merely left to tell the tale to the other Clallams up-Sound. The Clallam Indians believed that the Seeahtik Tribe had become extinct.

It is fifteen years since their tracks were last seen and recognized at the Brinnon River. Prior to that time, many Clallam Indians have met and talked with men of this strange tribe for the Seeahtiks talk the strange tongue of the Clallams, which is said to have originated from the bear tongue.

The Quinault Indians, however, claim that Fred Pope of the Quinault Tribe and George Hyasman of the Satsop Tribe were fishing about fifteen miles up the Quinault River in the month of September four years ago when they were visited by the Seeahtiks. The two Indians had caught a lot of steelhead trout, which they left in their canoe, and the Seeahtiks stole these.

Possesses Hypnotic Power

Henry Napoleon of the Clallam Tribe is the only Indian who was ever invited to the home of the Seeahtik Tribe. It was while Napoleon was visiting relatives on the British Columbia coast about thirty years ago (that would have made the year roughly 1895) that he met a Seeahtik while hunting.

The giant Indian then invited him to their home, which is in the very heart of the wilderness on Vancouver Island. Napoleon claims they live in a large cave. He was treated with every courtesy and told some of their secrets. He claims that the giant Indians made themselves invisible by strange medicine that they rub over their bodies and that they had great hypnotic powers and the gift of ventriloquism.

Some Indians claim that during the process of evolution when the Indian was changing from animal to man, the Seeahtik did not fully absorb the "Tamanaweis," or soul-power, and thus he became an anomaly in the process of evolution.

The Indians of the North-West are of the belief that the "mountain devils" found at Mt. St. Helens are the Seeahtik Indians as it is generally their custom to frighten persons who have displeased them by throwing rocks at them.

Ordinances and Resolutions: Sasquatch Protection

The continued insistence by witnesses and others that sasquatch/bigfoot do exist has prompted some government bodies to enact protective measures. The first was the Board of Commissioners of Skamania County, Washington, mainly brought about through the efforts of the noted bigfoot researcher Robert W. Morgan. It remained in effect for fifteen years and was then modified. I present hear both versions (older version first).

ORDINANCE NO. 69-01

BE IT HEREBY ORDAINED BY THE BOARD OF COUNTY COMMISSIONERS OF SKAMANIA COUNTY:

WHEREAS, there is evidence to indicate the possible existence in Skamania County of a nocturnal primate mammal variously described as an ape-like creature or a sub-species of Homo Sapien, and;
WHEREAS, both legend and purported recent sightings and spoor support this possibility, and
WHEREAS, this creature is generally and commonly known as a "Sasquatch", "Yeti", "Bigfoot", or "Giant Hairy Ape", and has resulted in an influx of scientific investigators as well as casual hunters, many armed with lethal weapons, and
WHEREAS, the absence of specific laws covering the taking of specimens encourages laxity in the use of firearms and other deadly devices and poses a clear and present threat to the safety and well-being of persons living or traveling within the boundaries of Skamania County as well as to the creatures themselves,

THEREFORE BE IT RESOLVED that any premeditated, willful and wanton slaying of such creature shall be deemed a felony punishable by a fine not to exceed Ten Thousand Dollars ($10,000) and/or imprisonment in the county jail for a period not to exceed Five (5) years.

BE IT FURTHER RESOLVED that the situation existing constitutes an emergency and as such this ordinance is effective immediately.
ADOPTED this 1st day of April, 1969.

BOARD OF COMMISSIONERS OF SKAMANIA COUNTY.

*****

ORDINANCE NO. 1984-2
PARTIALLY REPEALING AND AMENDING
ORDINANCE NO. 1969-01

WHEREAS, evidence continues to accumulate indicating the possible existence within Skamania County a nocturnal primate mammal variously described as an ape-like creature or sub-species of Homo Sapiens; and
WHEREAS, legend, purported recent findings, and spoor support this possibility; and
WHEREAS, this creature is generally and commonly known as "Sasquatch", "Yeti", "Bigfoot", or "Giant Hairy Ape", all of which terms may be hereinafter be used interchangeably; and
WHEREAS, publicity attendant upon such real or imagined findings and other evidence have resulted in an influx of scientific investigators as well as casual hunters, most of which are armed with lethal weapons; and
WHEREAS, the absence of specific national and state laws restricting the taking of specimens has created a dangerous state of affairs within this county in regard to firearms and other deadly devices used to hunt the Yeti and poses a clear and present danger to the safety of the well-being persons living or traveling within the boundaries of this county as well as to the Giant Hairy Apes themselves; and
WHEREAS, previous County Ordinance No. 1969-01 deemed the slaying of such creature to be a felony (punish-

able by 5 years in prison) and may have exceeded the jurisdictional authority of that Board of County Commissioners; now, therefore

BE IT HEREBY ORDAINED BY THE BOARD OF COUNTY COMMISSIONERS OF SKAMANIA COUNTY that a portion of Ordinance No. 1969-1, deeming the slaying of Bigfoot to be a felony punishable by five years in prison, is hereby repealed and in its stead the following sections are enacted:

SECTION 1. Sasquatch Refuge. The Sasquatch, Yeti, Bigfoot, or Giant Hairy Ape are declared to be endangered species of Skamania County and there is hereby created a Sasquatch Refuge, the boundaries of which shall be coextensive with the boundaries of Skamania County.
SECTION 2. Crime-Penalty. From and after the passage of this ordinance the premeditated, willful, or wanton slaying of Sasquatch shall be unlawful and shall be punishable as follows:
(a) If the actor is found to be guilty of such a crime with malice aforethought, such act shall be deemed a Gross Misdemeanor.
(b) If the act is found to be premeditated and willful or wanton but without malice aforethought, such act shall be deemed a Misdemeanor.
(c) A gross misdemeanor slaying of Sasquatch shall be punishable by 1 year in the county jail and a $1,000.00 fine, or both.
(d) The slaying of Sasquatch, which is deemed a misdemeanor, shall be punishable by a $500.00 fine and up to six months in the county jail, or both.
SECTION 3. Defense. In the prosecution and trial of any accused Sasquatch killer the fact that the actor is suffering from insane delusions, diminished capacity, or that the act was the product of a diseased mind, shall not be a defense.
SECTION 4. Humanoid/Anthropoid. Should the Skamania County Coroner determine any victim/creature to have

been humanoid the Prosecuting Attorney shall pursue the case under existing laws pertaining to homicide. Should the coroner determine the victim to be an anthropoid (ape-like creature) the Prosecuting Attorney shall proceed under the terms of this ordinance.
BE IT FURTHER ORDAINED that the situation existing constitutes an emergency and such this ordinance shall become effective immediately upon its' passage.
REVIEWED this 2nd day of April, 1984, and set for a public hearing on the 16th day of April, 1984, at 10:30 o'clock a.m.
BOARD OF COUNTY COMMISSIONERS
Skamania County, Washington
ORDNANCE NO. 1984-02 IS HEREBY DULY PASSED AND ADOPTED INTO LAW this 16th day of April 1984.

*****

The Whatcom County Resolution.

RESOLUTION NO. 92-043
DECLARING WHATCOM COUNTY A SASQUATCH PROTECTION AND REFUGE AREA

WHEREAS, legend, purported recent findings and spoor suggest that Bigfoot may exist; and
WHEREAS, if such a creature exists, it is inadequately protected and in danger of death or injury;

NOW, THEREFORE, BE IT RESOLVED by the Whatcom County Council that, Whatcom County is hereby declared a Sasquatch protection refuge area, and that all citizens are asked to recognize said status.

BE IT FURTHER RESOLVED, this resolution shall be effective immediately.

APPROVED this 9th day of June, 1991.

Sightings/Incidents Statistics (Compiled by C.L. Murphy)

This map/chart provides a reasonable idea of the distribution of sasquatch-related sightings and incidents in North America. The figures shown represent the number of what are thought to be "credible" reports up to 2003 (effectively 100 years). However, the true number of incidents would be much greater because most are not reported. It is estimated that the ratio is about 8:1. In other words, for every incident reported, eight are not reported. One would therefor multiply the figures shown on the chart by eight to arrive at what would probably be the true number.

To get an idea of how many incidents by state or province currently occur annually, the chart figures can be used to determine a percentage of the total, which is then applied to a current estimate of the annual number of incidents (reported and non-reported). At this time that estimate is 400 per year.

The figures shown on the chart total 2,557. If the question is, for example, what is the current annual number of incidents in Washington state based on the total for North America, the calculation would be 286/2557 x 400. This equates to forty-five. So it can be said that there are forty-five reported and non-reported incidents in Washington each year.

If one wishes to isolate the US or Canada in the 400-per-year figure, it is simply distributed on a percentage basis. Canada has 21 percent of the total, and the US 79 percent. This equates to Canada having eighty-four annual incidents and the US 316.

While all of this might be mathematically feasible, it cannot be used to reasonably estimate a sasquatch population in any area. One of the main factors that influences the figures is the number of people available to see sasquatch or find evidence of its presence. If, for example, Canada had the same population and distribution of people as the US, then there would likely be many more times the number of sasquatch-related incidents in Canada. Nevertheless, having said that, the other main factor is the number of sasquatch available to be seen in a given area. Naturally, the more sasquatch available, then the more likelihood that they will be sighted or will leave evidence. We see that British Columbia has the largest figure for North America, yet the people population in this province is far less than Washington, Oregon, or California. Consequently, it can be reasoned that BC simply has a greater sasquatch population than these states. How many sasquatch are there in North America? A good guess is 7,000 to 10,000 with about 80 percent living in (and staying in) Alaska, the Yukon Territory, British Columbia, Washington, Oregon, and California.

Noteworthy Artifacts & People

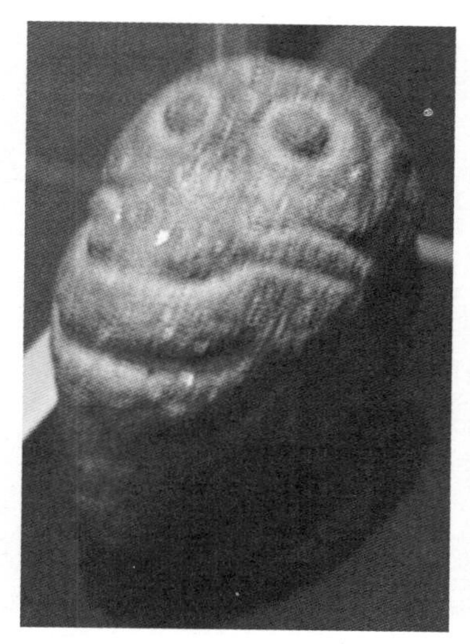

Probably the oldest known sculptured depiction of sasquatch are the stone heads, of which one is shown here. They were found in the Columbia River valley (Oregon and Washington) and date from about 1500 BC to 500 AD. Two anthropologists have stated that the heads have anthropoid (ape) features. Other stone heads show normal animals, so they rule out these associations. As there are no wild apes in North America, the sasquatch is considered a likely candidate.

The best petroglyph (image etched in rock) of a sasquatch is located at Painted Rock, California. The Native people refer to the creature as the "hairy man," and have a unique legend on how this creature and other forest creatures originated. The age of the image is not known, but would definitely exceed 500 years.
(Photo, Kathy Moskowitz Strain)

The best pictograph (rock painting) of a sasquatch is also located at Painted Rock, California, and the same information applies as that shown for the Painted Rock petroglyph on the previous page. The full rock image also shows the hairy man's wife and child, although not as discernible as the hairy man himself. (Photo, Kathy Moskowitz Strain; drawing, Brenden Bannon)

Possibly the oldest published image of a non-human primate published in North America appeared on the front cover of a 1785 edition of *Bickerstaff's Boston Almanac*. It is said to show an "ape" from Africa. It is based on an earlier illustration by the pioneering primatologist Edward Tyson. It has a strange similarity to the sasquatch, and if this was the intent, then it would be the earliest known non-Native depiction of the creature. (Image courtesy of Dr. Brian Regal,Kean University, New Jersey)

Probably the oldest Native drawings of sasquatch. They were both created by David Cusick (d. 1840), a Tuscarora Native. The first, created in 1820, shows a Native woman parching acorns with a surprised "cannibal monster" watching her. The creature was frightened away because he thought the woman was eating red hot coals.

The second shows "Stonish Giants" chasing Native people. The giants, it is said, started to overrun the country in about 242 AD.They were so ravenous that they devoured the people in almost every town. With the help of the "Holder of the Heavens" the giants were defeated and forced to seek asylum in the regions of the north.

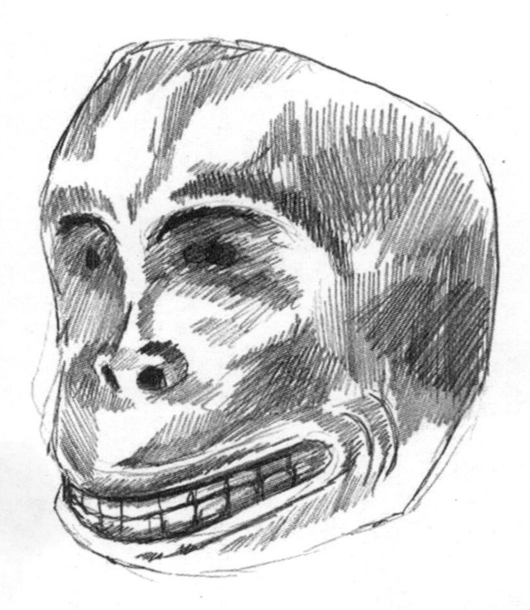

The best Native sasquatch mask is this Tsimshian mask found in British Columbia in 1917. The actual mask was probably created in about 1850. There is no doubt that it shows an ape-like creature. The Tsimshian people state that they have a belief in "mountain monkeys." (Artwork by Peter Travers)

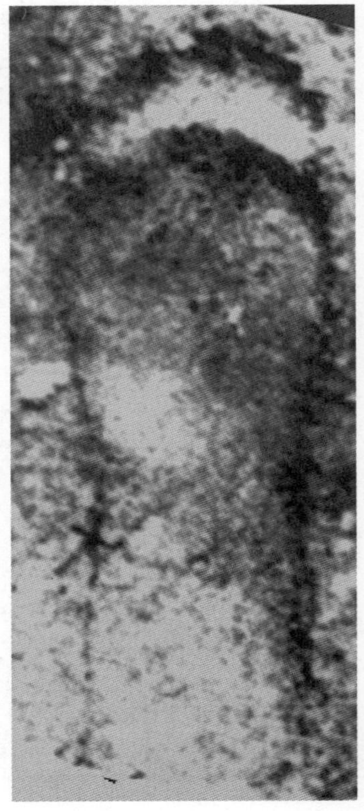

The oldest photograph of a sasquatch footprint was taken in October 1930 at a location about two miles below Spirit Lake, Mt. St. Helens, Washington. The print was sixteen inches long. The photo came to light in August 1963, when Marge Davenport of the *Oregon Journal* was doing research on the sasquatch at the Spirit Lake Ranger Station.

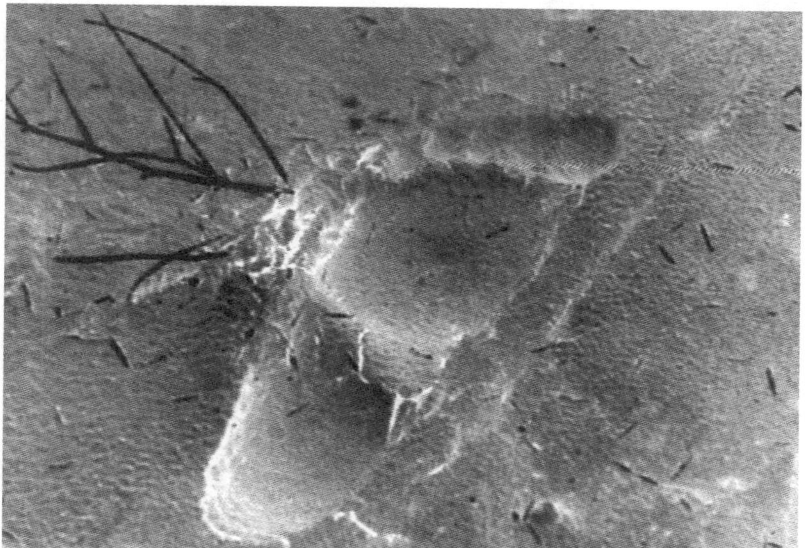

The best photograph of a sasquatch footprint was taken by Lyle Laverty at the Patterson/Gimlin film site on October 23, 1967. The first images shows the print as it appears in the photo. The second image is the same photo *inverted* to eliminate the optical illusion associated with some photos (prints appear raised rather than impressed). As can be seen, the creature stepped on a twig. The break in the center of the print has significant scientific meaning as to the nature of the creature's foot. The print length was 14.5 inches.

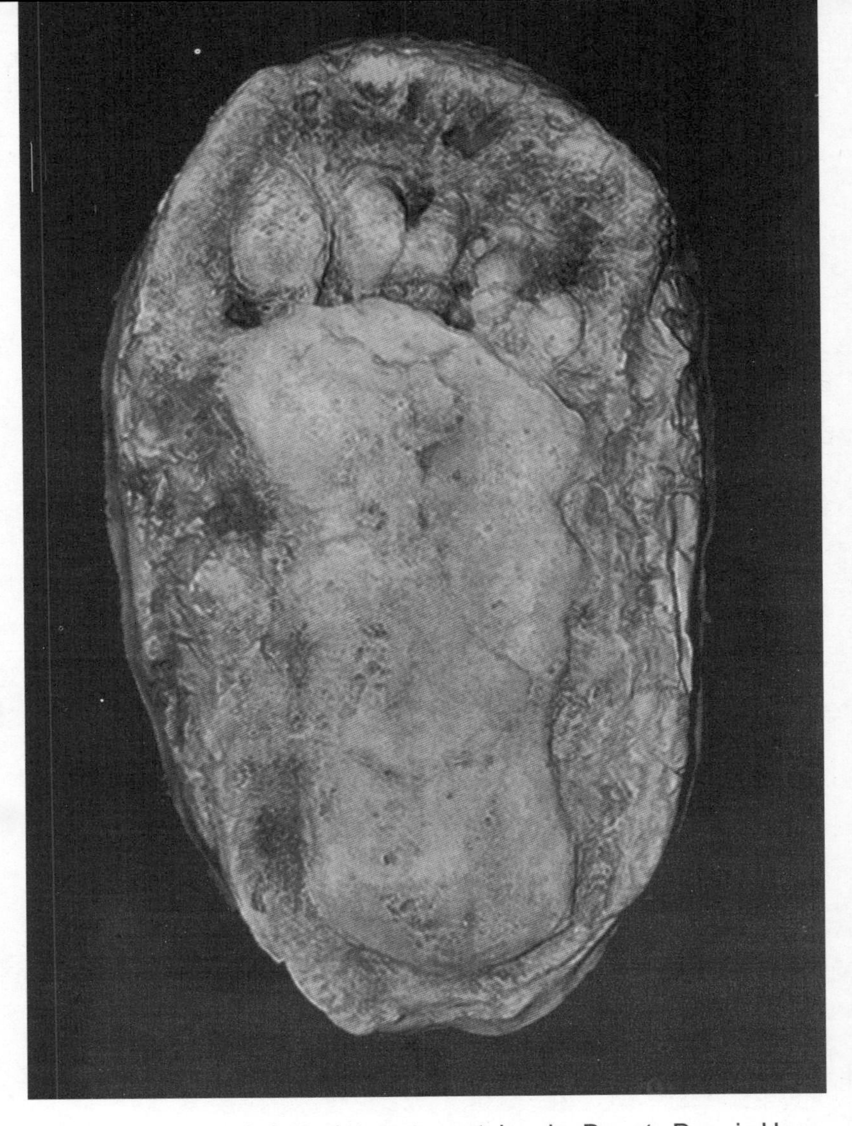

The best sasquatch footprint cast was taken by Deputy Dennis Heryford at Abbott Hill, Washington in 1982. Many prints were found, along with half-prints that indicated a fully flexible foot. The prints were about fifteen inches long. The foot configuration is different from that of the creature in the Patterson/Gimlin (Laverty photo) and perhaps this is because that creature was female and this creature was male. For certain, human female feet are much "neater" than human male feet, and I would suspect that the same applies to sasquatch. This excellent reproduction of the Heryford cast was made by Rick Noll. It is an astounding artifact.

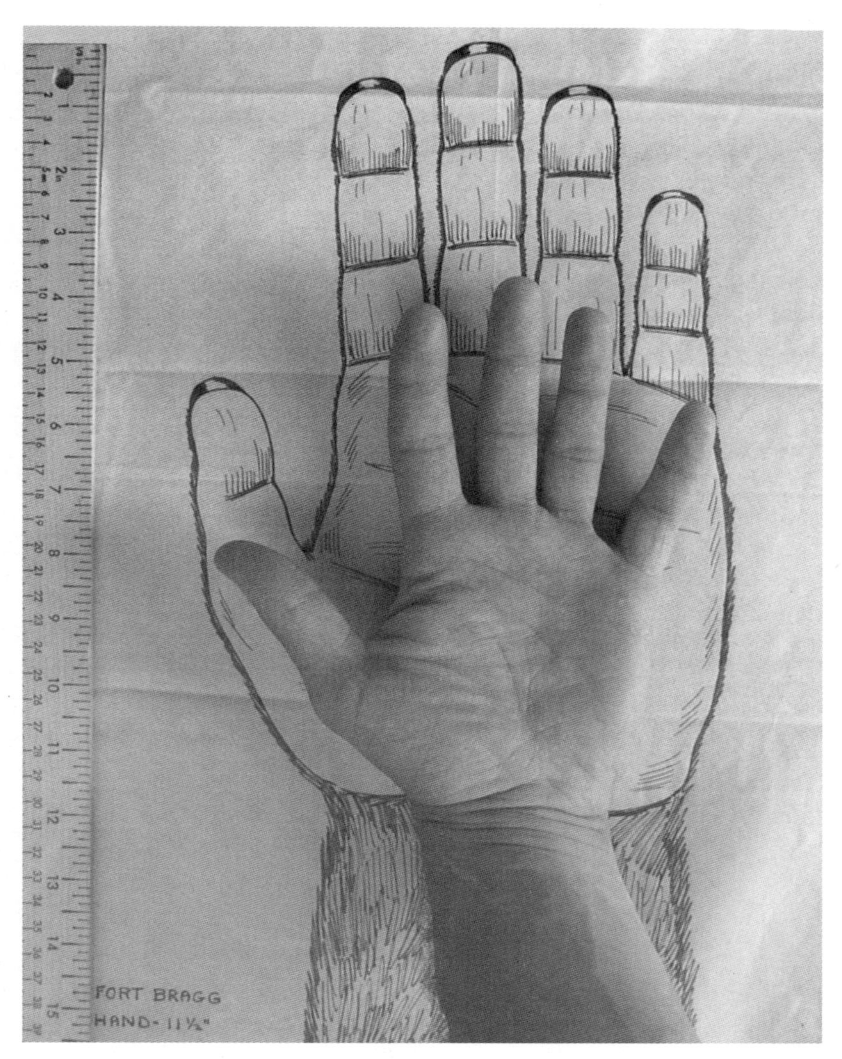

The best tracing of a sasquatch hand print was made at Fort Bragg, California in February 1962. The creature was observed in the backyard of a house owned by Mr. and Mrs. Bud Jenkins. It approached the back door of the house, which was open. Mrs. Jenkins got to within two to four inches of closing the door when the creature pushed back. Mrs. Jenkins kept pushing, and after a few moments the creature simply walked away, leaving a dirty hand print. The human hand shown for comparison is of a six-foot-tall male, 200 pounds. The hand print tracing (base of palm to longest finger tip) is 11.5 inches long.

The first photograph of a sasquatch was probably taken by Zack Hamilton in the 1950s. He took the film to a shop in San Francisco to have it developed. He told the clerk that he was a woodsman, and that he had been stalked by some sort of hairy monster while in the Three Sisters area, Oregon. The film roll showed shots of the creature, but Zack never returned to pick up his photographs.

In 1965, *The San Francisco Chronicle* carried an article about the search for the sasquatch, which led the camera shop clerk, Dick Russell (by then the assistant manager), to recall Zack Hamilton and the photographs that had never been claimed. Russell had looked at the photos about three years before and saw that they showed something very odd. He called *The San Francisco Chronicle,* and on December 7, 1965 they featured an article about the Zack Hamilton event. Russell is quoted in this article as saying, when he first examined the photos, "I got prickly all over when I realized they were the pictures the old timer said he had taken in the brush. I never saw anything like them." *The Chronicle* article showed only one of the photos, seen here. The rest of the photos were never published and have not come to light.

The first reasonable portrait of a sasquatch was created by Chris Murphy in 1996. An image from the Patterson/Gimlin film was enlarged and photocopied. Chris then worked on the image with pastels, enhancing what details he could see and guessing details that were not available. The creature's mouth was closed to give it a more "pleasing" appearance and a decent chin was added. Many posters were created and the image has been used on the covers of several books, and as an illustration in books and magazines. It is likely the most publicized portrait (artwork) to date.

The first sasquatch image by a professional artist based on the creature seen in the Patterson/Gimlin film was by Peter Travers in 2002. Travers' exceptional work gives us a good idea of what sasquatch may actually look like. His skillful blending of human-like and ape-like features is very close to what many sasquatch witnesses say they have observed.

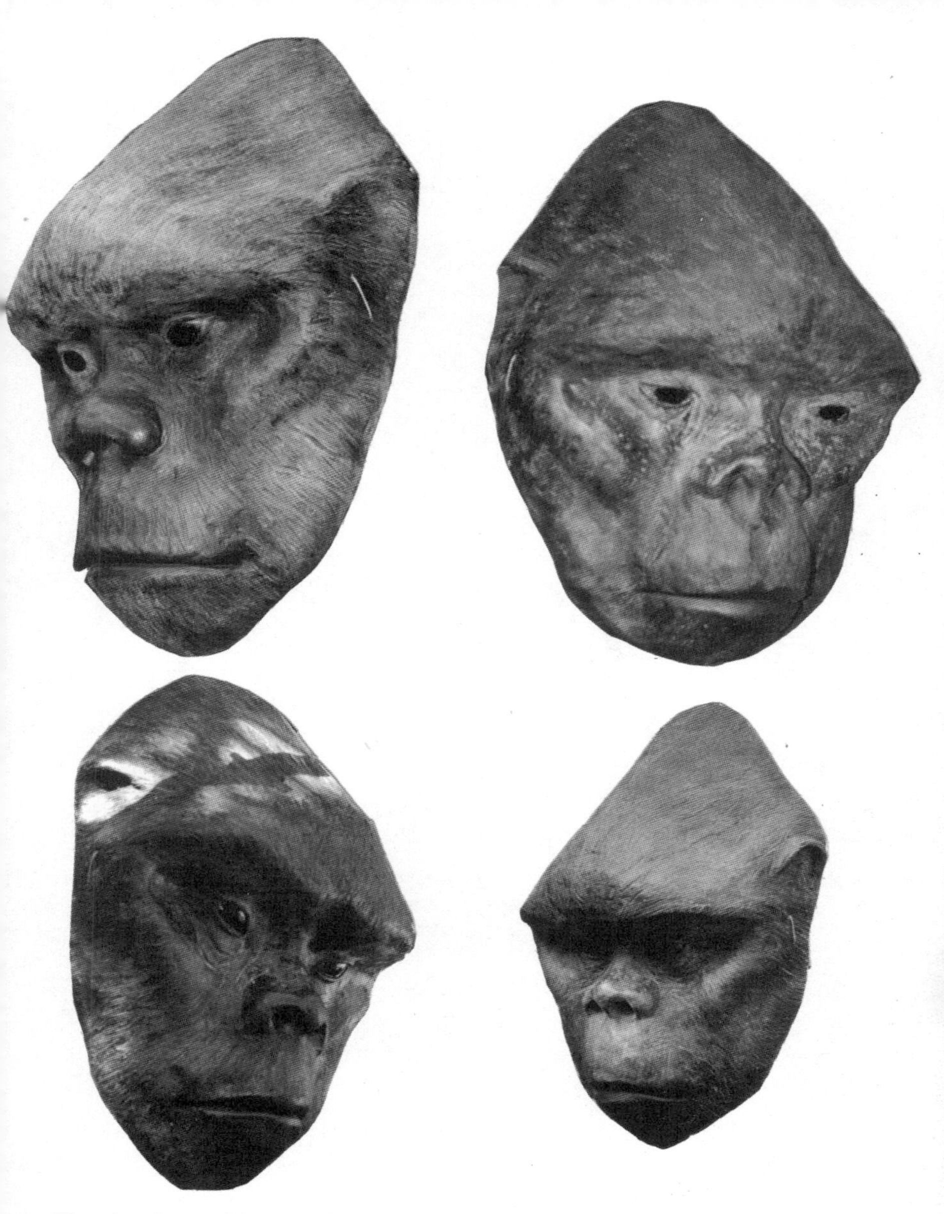

The best non-Native sasquatch head sculptures were created by Penny Birnam, a Vancouver, B.C., artist in 2003. They were created for a sasquatch exhibit at the Vancouver Museum in 2004/5. Each head has different facial features, as it is believed sasquatch, like humans, would have individuality. The heads vary in size. The first head (upper left) is 18.5 inches high, however these images are not proportional.

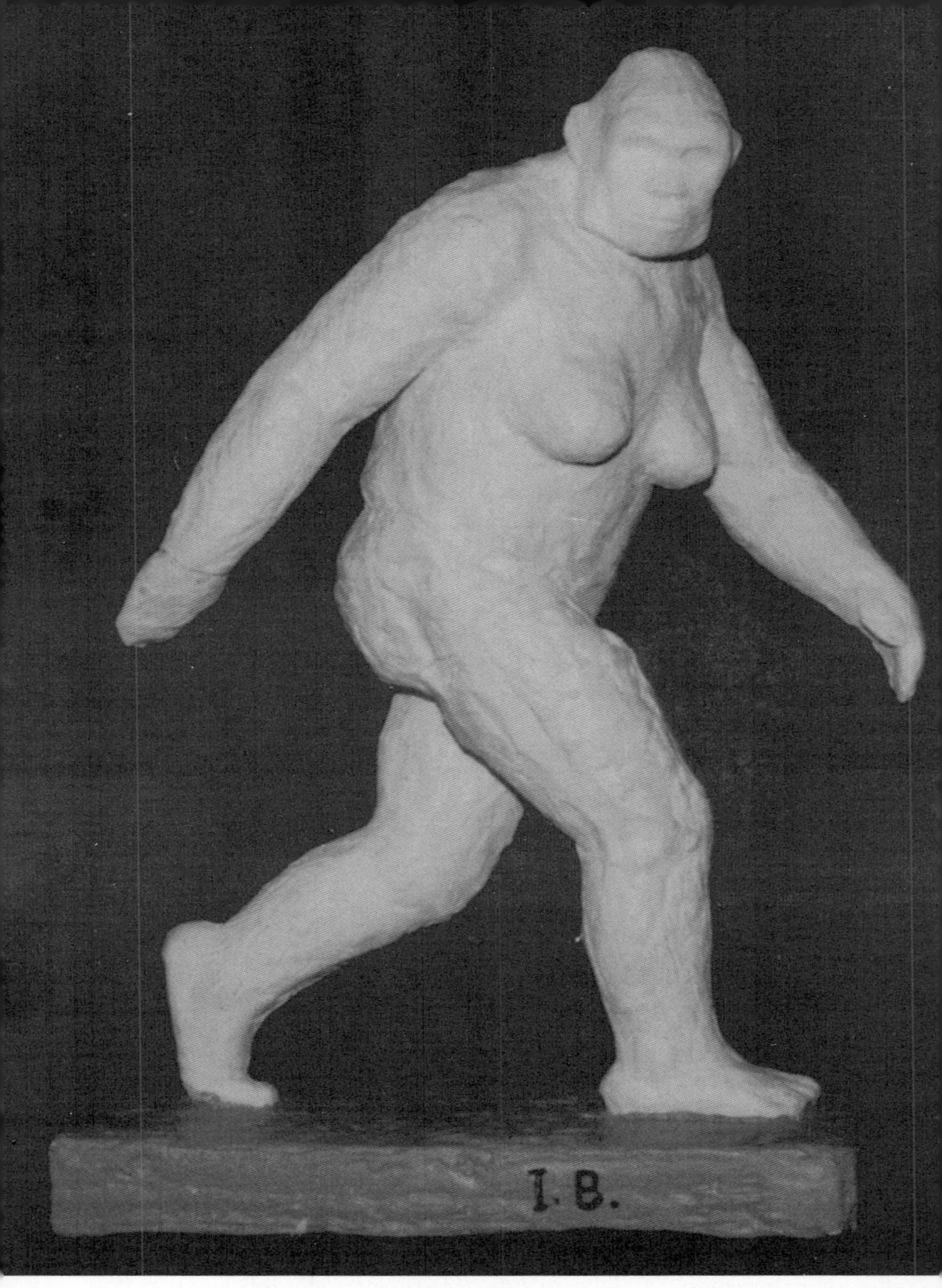

The first sculpture of the creature seen in the Patterson/Gimlin film was created by Igor Bourtsev, a Russian hominologist, in the early 1970s. He made several copies. The sculpture shown here was gifted to René Dahinden after his visit to Russia in 1971.

The only object for which we have photographic evidence that a sasquatch touched is this wood fragment. The creature in the Patterson/Gimlin film sort of stepped on it as she progressed along the gravel sand bar at Bluff Creek. In the film, the fragment can be seen to move slightly, so we are sure contact occurred. When René Dahinden visited the film site in 1971, he saw that the fragment was still in place. He took it home with him to Richmond, BC, and many years later (early 1990s) it was used as a cursory indicator to determine the creature's height, which came out as about seven feet tall. The fragment is about 26.25 inches long, 3.25 to 3.5 inches high and 1.25 to 1.50 inches wide (height and width vary). It weighs about two pounds. Studies indicate that it can be reasonably confirmed that the fragment was the same as that seen in the film.

The first (and only) government-issued postage to depict a sasquatch was released by the Canadian government in 1990. It was part of the Canadian Folklore series entitled Canada's Legendary Creatures. The Patterson/Gimlin film was instrumental in the design of the stamp, and the government publication on the stamp series references the film and shows one of the film frames.
(© Canada Post Corporation, 1990. Reproduced with permission)

The first and greatest of the main 20th century sasquatch researchers was Bob Titmus (d. 1997). His interest, and long search for the creature began in 1958 when he taught Jerry Crew how to make a plaster cast of one of the prints he found in the Bluff Creek, California area. Titmus had two sightings of the creature. He investigated many footprint discoveries and took excellent casts.

The most referenced sasquatch-related totem pole is this Kwakiutl pole that shows D'sonoqua, the cannibal woman, with her child. Some Native legend has it that the sasquatch were fearsome cannibals. It is likely the size and look of the creature resulted in this myth.

This most elaborate native sasquatch mask is this Kwakiutl "transformation" mask. It is used to portray different characters in Native dance stories. It is operated by pulleys. I do not know its age, but believe it has been used for a considerable time.

The four main high-profile sasquatch researchers to date. They are (left to right) John Green, Dr. Grover Krantz (d. 2002), René Dahinden (d. 2001), and Peter Byrne. Known as The Four Horsemen of Sasquatchery, these individuals spearheaded sasquatch research for up to fifty years (Green, Dahinden, and Byrne since 1958; Krantz since 1963). Although they differed greatly in personality and opinion, they all held a common belief in the reality of the creature, and made major contributions in the field of sasquatch studies. All four authored books.

Paranormal Aspects (by C.L. Murphy)

Although scientists in general and "normalist" bigfoot researchers have absolutely no time for any talk of the paranormal, it remains a contentious and much-debated issue.

Remarkably, the contention that the creature was paranormal or supernatural is far older than science as we know it. This was, and continues to be, the belief of many of North America's aboriginal people whose presence on this continent dates back to about 8,000 BC. Science, of course, rationalizes this fact by pointing to the many other creatures in Native myth and legend, which also have paranormal aspects. In other words, if you believe what is said about these other creatures as being mythological, then you must concede the same for bigfoot.

However, there is an importance difference. Bigfoot is the only Native "mythological" creature that has been seen, and continues to be seen, by thousands of non-Native people. It is also the only one that appears to leave physical evidence (footprints, handprints and so forth).

Naturally most non-Native people don't believe in mythology, so the creature is given "natural status." As such, it is stripped of its paranormal trailings and sought after as one would seek any other natural animal (with guns, cameras, and all manner of electronic devices).

Given the number of years (about 200) since Europeans reported seeing the creature, the amount of physical evidence and photographs we have that support the creature's existence as a natural animal is, in a word, pitiful.

There can be no doubt that this fact has resulted in convincing some people that the creature is of a paranormal nature as contended by Native people. In other words, it can disappear at will and some how manage to remain "unconfirmed" as to its reality.

Although one can sort of dismiss this, it is not the end of the story. Many people state they have had actual bigfoot-related experiences that are far beyond the boundaries of modern science and the laws of physics. They are of the belief that the creature is a "dimension traveler." While it is in this (our) dimension, it is just as physical as you or I. It eats, sleeps, defecates, and leaves physical evidence in the form of footprints, handprints, hair and so forth. It can then go to another dimension (presumably where it normally resides), which we have not yet learned to access.

Indeed, many of these witnesses claim to have telepathic communications with the creatures, which they insist are actually people of two different types; one type is far more ape-like than the other. I will mention here that this anomaly does occur in regular sasquatch witness accounts.

The message the bigfoot or sasquatch people have imparted is that they are very concerned with our present treatment of the planet earth. They have a major concern with nuclear war possibilities, which, it appears, will also impact their home dimension.

Be that what it may, there is also a further complication (depending on how you look at it), which implies that bigfoot people are connected with alien beings. A considerable number of UFO sightings have occurred at the same time as bigfoot sightings or activities, so a connection was imminent. One thought is that bigfoot were brought to earth by aliens. This being the case, I can only conclude that bigfoot's ability to slip into another dimension is a defense tactic also applicable to aliens. Certainly UFOs do effectively "disappear."

There are, of course, many questions that can be asked on all of this. The most obvious is, why have not bigfoot, or indeed aliens, made their existence fully known to humans? In other words, proclaim their existence and message on the steps of the U.S. Capitol in Washington, DC. As it is, only certain people appear to have had this privilege.

The only answer I can offer is that they can't do so, at least not at this time. There appears to be some governing factor that limits full disclosure.

I have provided this material to simply acknowledge that the "sasquatch and the paranormal" is now both a Native and non-Native part of the creature's history.

Chapter 7

Stomping Grounds

The Pacific Northwest has many stops of interest for the bigfoot enthusiast. Provided here is information on some of them that I have visited or am aware of.

SEATTLE MUSEUM OF THE MYSTERIES
Address: 623 Broadway E. Seattle, Washington
Phone: 206 328 6499

Description: This museum features many bigfoot-related artifacts and historical information on the creature. It is also a paranormal science museum exploring northwest legends and lore, e.g.,UFOs, ghosts. Many unusual artifacts are on display and visitors can take part in fun activities.

(Photo, Steve Mandich)

INDEX BIGFOOT STATUE

Location: Along Highway 2 at Mile Marker 36, east of Monroe, Washington, in the small town of Index (Highway 2, along the Cascade Mountain pass).

Description: The town is most known for being a location used in the 1987 film *Harry and the Hendersons.* A giant wooden bigfoot carving that resembles "Harry" stands on the outskirts of the town. Visitors come here to relax at the small park, enjoy the magnificent and majestic mountain scenery, have their picture taken with "Harry," and partake in a good cup of coffee at the adjacent roadside espresso stand. (Many blackberry bushes abound in the summer months; bring a bucket!)

(Photo, Steve Mandich)

COLVILLE SASQUATCH SCULPTURE

Location: Colville Indian Reservation, eastern Washington State.

Description: A shock to see along Disautel Pass between Omak and Nespelem on the Colville Indian Reservation is an incredible twelve-foot sasquatch that appears to step out from the trees above the highway. The artist, Virgil "Smoker" Marchand, who has many more plans for the area, created the intriguing sculpture.

(Photo, Michael C. Scott)

KID VALLEY BIGFOOT STATUE

Location: Kid Valley, Washington, A roadside attraction with a bigfoot statue, the buried A-frame, and North Fork Survivors gift shop.

Directions: Take Interstate 5 exit 49, head east on Highway 504 (Spirit Lake Memorial Highway) twenty miles. About a mile east of the intersection of Highway 505 at mile post 20.5.

Description: About eight-and-a-half hours after Mt. St. Helens blew her stack on May 18, 1980, a river of volcanic mud and ash arrived in Kid Valley, some twenty-five miles distant, destroying the homes in this forested hollow along Highway 504. The destruction included a newly built A-frame house, which filled with two hundred tons of silt, mud, water, and ash. The A-frame remains, and you can walk through the ground floor (now below ground) since the mud-flow raised the elevation by a few feet. You can also see the twenty-eight-foot-tall bigfoot statue made of concrete, with detailed hair/fur and a "folklorish" grin. The adjacent gift shop offers numerous curios and specialty items.

APE CAVE LAVA TUBE

Location: In the hills, south of Mt. St. Helens. From Interstate 5, go 40 miles east of exit 21to Woodland, Washington. A large parking lot is provided.

Description: Ape Cave is a 2000-year-old lava tube, a giant underground passageway in the middle of sasquatch country. It was scoured out by a river of lava that poured down from the volcano above, resulting in a cave more than two miles long. Ape Cave is the longest lava tube in the western hemisphere, and one of the biggest in the world. At least a few bigfoot enthusiasts think Ape Cave is the beginning of an underground network of tunnels used by bigfoot to cross the continent.

For visitors, there are stairs that lead deep down into the cave, with interpretive information at the entrance. But stepping down into a deep hole in sasquatch country still feels a little eerie. As the sunlight disappears behind you and you step into a cold, wet, subterranean world, the only sounds you hear are the echo of your footsteps and trickling water. While your eyes adjust to the dark, you begin to feel very much alone. But are you? You begin to feel as if someone is watching you. And maybe someone is.

USFS Photo by Gifford Pinchot National Forest

BIGFOOT TRAP

Location: Take Highway 238 from Jacksonville, Oregon, onto upper Applegate Road toward the Applegate Dam. A pull-off alongside the road is used for parking. Hike just over half an hour to an abandoned miner's claim and follow a dirt trail to the trap.

Description: Applegate Valley is nestled in the Rogue River National Forest, and has been an area of much bigfoot activity in the past. It is also well known as the site of the world's only bigfoot trap. In 1974 a group of locals, who called themselves The North American Wildlife Research Group, built the trap in an attempt to capture, tranquilize, attach a transmitter, and then release one of the elusive creatures.

The trap, a ten-by-ten-foot square, heavy, wooden box is reinforced with steel bolts and plates, and is anchored to thick poles. During its use, a rabbit carcass would be hung from the ceiling of the contraption; the theory was that once the bigfoot grabbed for the rabbit the trap door would slam shut, an alarm would sound and, *voila!*—you've got yourself a bigfoot! Well built and strong, the trap still stands today as the day it was constructed. There is a tremendous amount of interest in bigfoot, so much so that the forest rangers put up an informative sign at the trap site.

WAX BIGFOOT, Mariner Square Wax Works Museum
Address: 250 SW Bay Boulevard
Newport, Oregon
Phone: 541 265 2206
Website: www.marinersquare.com

Description: The wax figure of a bigfoot at the Mariner Square Wax Works Museum in Newport is said by many eyewitness to resemble what they have seen. Not only will you see bigfoot at the wax museum, there are also hundreds of life-like Hollywood legends and sci-fi creatures galore.

Situated along the beautiful shoreline of the Oregon Coast, Mariner Square has long been a favorite destination for travelers from near and far. The bustling bay front is where all the action is, lined with delectable restaurants, galleries, fisherman's markets, a picturesque marina, sandy beaches, and lighthouses. For more fun, be sure to check out the exciting and bizarre Ripley's Believe it or Not and the Undersea Gardens.

BIGFOOT'S PUB 'n GRUB
Address: 2427 S. Roosevelt Dr. Seaside, Oregon
Phone: 503 738 7009

Description: While visiting the sandy beaches along the beautiful Oregon Coast, be sure to stop in at Bigfoot's Pub 'n Grub for some delicious food. Order from the bigfoot-related menu and see the bigfoot statue inside the pub. Ask the owner about bigfoot. He has many stories to tell.

WILLOW CREEK–CHINA FLAT MUSEUM

Location: Trinity Highway (Highway 299), Willow Creek, California (near the intersection of route 299 and route 96).

Phone: 530 629 2653 (call for hours)

Description: A place of nostalgia, this museum displays "the good old days" of Trinity and Humboldt Counties, as well as Indian artifacts of the various tribes from the surrounding area. But the big attraction of course is the bigfoot exhibit containing generous donations of items, including footprint casts, among which are the original casts taken by Bob Titmus. There are many bigfoot photos, maps, and other bigfoot evidence exhibited in an extension built specifically to house this collection. The parking lot guarded is by a twenty-five-foot redwood bigfoot sculpture. A must-see for bigfoot enthusiasts.

The town of Willow Creek itself is the self proclaimed "Gateway to Bigfoot Country." A huge redwood carving of bigfoot by Jim McClarin stands in the center of town. Every Labor Day weekend Willow Creek plays host to Bigfoot Days, the largest celebration of the creature anywhere. Food, parades, arts, crafts, and a party with much fun await you. There are also other events and parties throughout the summer months.

Willow Creek is a rugged mountain community nestled in the heart of the Six Rivers National Forest. This area of California is located in the Trinity/Shasta/Cascade region. Outdoor recreational enthusiasts love Willow Creek, where the most popular activities are camping, fishing, and rafting, each of which are second-to-none. Other activities nearby include swimming, hiking, snow skiing, horseback riding, rock climbing, off-road racing, golf, birding, and wildlife viewing. It's also an incredible area for scenic drives. Organic gardens and vineries abound and welcome travelers to try their wares. Bigfoot sightings also abound; in fact, it was here in 1967, in a nearby canyon of Bluff Creek, where the Patterson/Gimlin film was shot, showing what appears to be an adult bigfoot walking along the creekside.

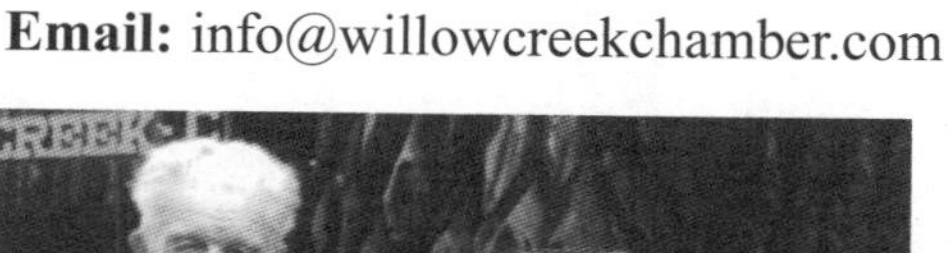

Contact: Willow Creek Chamber of Commerce 800 628 5156
Email: info@willowcreekchamber.com

Al Hodgson (right), curator of the Willow Creek–China Flat Museum, is seen here with John Green. Al was the first person Roger Patterson and Bob Gimlin contacted after they filmed a bigfoot at Bluff Creek on October 20, 1967. (Photo, Daniel Perez)

PATTERSON/GIMLIN FILM SITE
Location: On Bluff Creek, California, in the Six Rivers National Forest (a highly remote area).

Description: For the physically fit and brave at heart who want an adventurous hike. It was here on October 20th, 1967 that Roger Patterson and Bob Gimlin took the most controversial film footage of a sasquatch to date: the Patterson/Gimlin film. A plaque has been erected on a river-rock monument at the site where the creature was spotted and the footage taken. Memorial donated by researchers Sean Fries, M.K. Davis, M. Gross, and J. Heirs of the Bigfoot Research Project.

Directions: To get to the site, take Highway 299 in Willow Creek California. Go north on Highway 96 toward the town of Orleans. Just as you come into Orleans, on your left there is a road called Eyesee Road; turn left there. Go seventeen miles up to Forest Service Road 12N12, and turn left again. (There is a sign that says: Louse Camp, 9 Miles.)

Travel to Forest Service Road 12N13 and turn right, (should be the first right you can make). From there you need to stop and set your trip meter and go exactly 3.6 miles. There, to your left, you will see a little side road that goes uphill slightly. Take it, and as you come over the hill, the road splits off into two sepa-

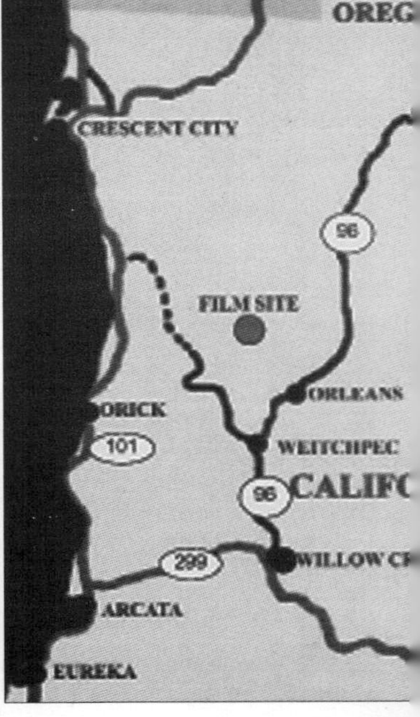

rate roads. Stay to the right; it's called K Spar. Follow it down to the bottom or as close as you can get.

There are sometimes rock slides, and if this is the case you will have to walk the rest of the way in. When you reach the bottom it opens up and you will see a primitive campsite. It's the only campsite there and this is the place to be. From this site, the creek is just to your right. You need to go down stream about 500 yards where it opens up to the Patterson/Gimlin site—just look for the monument. (Please be respectful of the area and site).

Guided Tours
Phone: 530 623 6237
Email: friessean@aol.com
Address: Sean Fries, P.O. Box 3286, Weaverville, California, 96093
This is a guided tour of the Patterson/Gimlin area and film site. Learn how to look for signs of bigfoot, how to collect evidence, and much more. Whole-day trips, or overnight trips available.

Chris Murphy is seen here at the film site as it appeared in 2003. Only a portion of the actual site remains. About two-thirds of it is now a rocky gorge as a result of Bluff Creek overflowing and gouging out its banks. The creek is fed by numerous mountains, so in heavy rains it becomes a raging torrent destroying everything in its path.

CAPRITAURUS BIGFOOT DISCOVERY MUSEUM

Location: 5497 Highway 9, Felton, California. Located in the San Lorenzo Valley in the Santa Cruz Mountains of Northern California, in the heart of the coastal redwoods, just six miles above the beach/boardwalk of Santa Cruz, next to Henry Cowell Redwoods State Park and the Roaring Camp Railroads.
Phone: 831 335 4478
Email: info@Bigfootdiscoveryproject.com
Website: www.Bigfootdiscoveryproject.com

Description: More than just another roadside attraction, it's also a bigfoot collection, library and research group, a center for scientific exploration, a wildlife awareness and conservation project, a local history project, a artist's studio and gallery, a gift shop, bookstore, and much more.

The museum curator, Michael Rugg (seen in the above photograph), provides the following information:

> This project is based on the knowledge that people of all demographics report sightings and encounters with bigfoot (unknown or mystery bipedal primates) including deputy sheriffs, forest rangers, loggers, hunters, and wildlife biol-

ogists. Something leaves apparently authentic, large human-like tracks in the forest. Primate-like hairs have been collected that match each other but no other known primate. Recent high-tech scrutiny of the Patterson/Gimlin film has added details to authenticate it; and it has never been debunked, except by hoaxers, liars, and unscrupulous journalists.

The Bigfoot Discovery Project (BDP) accepts the Patterson/Gimlin film of October 20, 1967 as THE type specimen for the northwestern bigfoot or sasquatch. The BDP will add to the dialogue on the implications of the impending "discovery of bigfoot" by western science and the general public, while offering hypotheses as to bigfoot's place in the biological and social order. The BDP will seek to facilitate the discovery of bigfoot by sharing information, and attempting to "flesh out" the hairy biped's dossier.

This will be done against a backdrop of local history, as this area was once (and apparently still is) home to bigfoot, as are all the forests of the Pacific coast and beyond. The new CapriTaurus complex, once complete, will include exhibits of local history, unidentified hairy bipeds, and folk music. It will house a research library of books, periodicals, and tapes on subjects ranging from ABSMs to UFOs, alongside art, handcrafts, and collectables.

We hope that a visit here will "edu-tain" you, and help to open your eyes and mind to things you may not have thought about before. Our collection includes exhibits of local history, tied in with local bigfoot sightings, popular culture as it relates to the public view of bigfoot, and actual evidence in the form of plaster foot and hand prints, along with a detailed exhibit on the Patterson/Gimlin film. There are exhibits inside the main building as well as in the rear, a nocturnal diorama featuring bigfoot and an outdoor audio-video "theatre/cafe" area where you can see bigfoot documentaries of your choosing.

Our mission: to help the general public in their discovery of bigfoot; create a library and study center for mysteries and wonders of science; teach children reverence for nature and other living things; seek physical evidence for bigfoot. Apparently they are all around us, right here in the San Lorenzo Valley, in mountains that overlook the very heart of Silicon Valley!

In the two years our doors have been open, we've received well over 100 reports of bigfoot, UFOs, ghosts, little grey men, and other unknown animals. It's been a veritable smorgasbord of "paranormal" anecdotes. We have been introduced to the reality of "backyard bigfoot," and have begun to recognize the most likely habitats and corridors for the "locals"—the forest giants of the Santa Cruz Mountains.

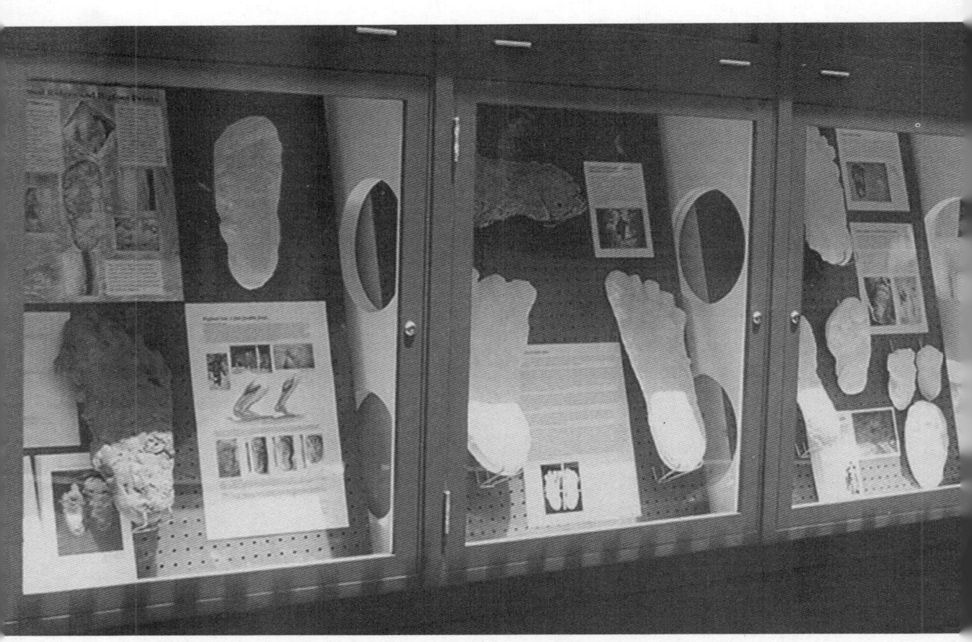

Artifacts on display at the Bigfoot Discovery Museum.

THE BIGFOOT LODGE
Address: 1750 Polk Street
San Francisco, California
Phone: 415 440 2355

Description: See the nine-foot, six-inch resin bigfoot! It's billed as: "It's big, it's hairy, and it will stomp your ass!" The rustic log cabin, pine-siding cocktail lounge was built around him. This is a fun, kitschy, campy, rock-and-roll bar. A simulated forest fire burns above the bar where a little bigfoot gets smoked out. The lodge is open nightly from 5:00 p.m. to 2:00 a.m.

LEGEND OF BIGFOOT SHOP
Address: 2500 Highway 101, Garberville, California
Phone: 707 247 3332

Description: This bright and colorful souvenir shop is chalk full of fun stuff—bigfoot creations, wood carvings of bigfoot and other creatures, postcards, kites, toys, souvenirs, hats, T-shirts, collectables, and much, much more.

(Photo, Steve Mandich)

SASQUATCH CROSSING LODGE, Canada
Location: Mile 147, Alaska Highway, Pink Mountain, BC.
Phone: 250 772 3220

Description: Your highway stop between Dawson Creek and Ft. Nelson. The Rocky Mountains can be seen stretching across the western horizon. Hunt and fish in the beautiful Pink Mountain Valley. A tall sasquatch statue stands guard in front of the store, which serves gas, and has gifts and souvenirs. The RV Park has fifteen sites and a dump station. Tent sites are also available, as well as comfortable and clean cabins, and an eleven-room motel with open camp services. Enjoy home style cooking in the restaurant. Open year round.

HARRISON SASQUATCH CARVING, Canada
Address: The Springs Campground
670 Hot Springs Road
Harrison Hot Springs, BC
Phone: 604 794 9907

Description: The sculpture is on the roadside at the entrance to the campground. It has long been a main attraction at Harrison for tourist photographs. Seen here "measuring up" is Tony Healy from Australia.

SASQUATCH TOURS, Canada
Address: 5061 Chehalis Road, Comp. 18, Agassiz, BC, V0M 1A1
Phone: Toll free: 877 796 1221
Email: info@sasquatchtours.com
Website: www.sasquatchtours.com

Description: "A First Nations cultural experience." Operates enriching and memorable aboriginal cruises, interpretative programs, and workshops within the beautiful coast mountains of southwestern British Columbia, on pristine Harrison Lake and Harrison River.

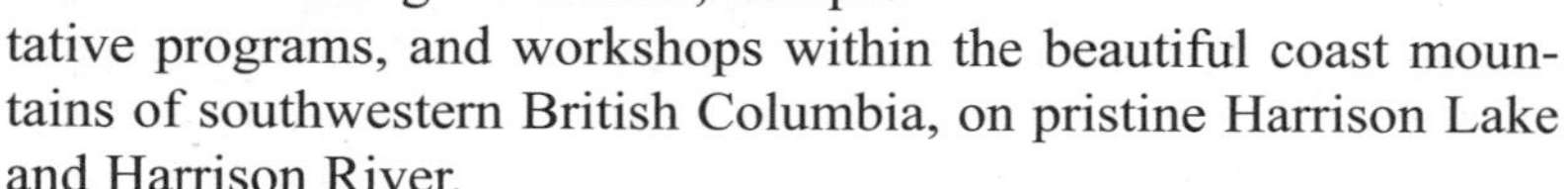

Kelsey Charlie *(Tixweltel)* and his daughter, Angela Raven *(Selyaltenawt),* in 2005. The Charlie family runs Sasquatch Tours. Kelsey is holding a drum that shows the Chehalis Band logo (inspired by the creature seen in the Patterson/Gimlin film). There have been many sasquatch sightings in the Chehalis reserve area. Kelsey has stated that he saw (with another witness) a mother with a young one, and his father and grandfather both said they had sightings.

THE SASQUATCH INN, Canada
Address: 46001 Lougheed Highway, Harrison Mills, BC
Phone: 604 796 2730 (Canada & USA); 604 796 2730 (International)
Email: sasquatchinn@shaw.ca
Website: www.sasquatchinn.ca/

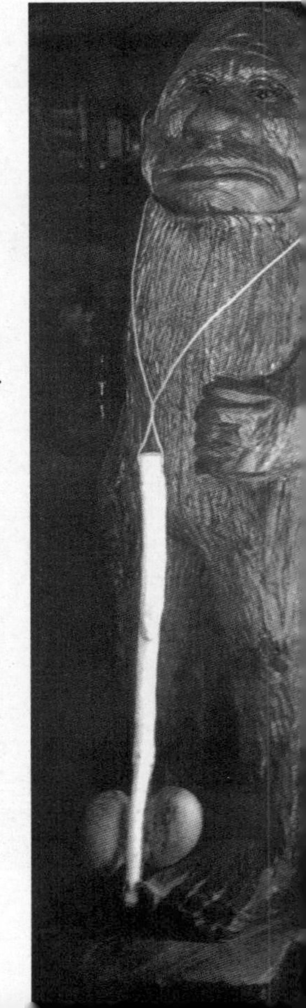

Description: The Sasquatch Inn is a famous small country inn located at Harrison Mills, British Columbia, Canada. Located right on the turn off that goes to the Chelalis Indian Reservation, it has for years been a favorite with those in the sasquatch fraternity. A sasquatch carving stands right inside the restaurant door. Accommodations include fourteen guest rooms, restaurant, pub, and a liquor store. The restaurant has an outstanding reputation for great food and great service. Offering a large selection of fine dining, catering both to vegetarians and meat lovers alike, some of the popular bigfoot items listed on the menu are: The Legendary Sasquatch Burger, the Sasquatch Sub Sandwich, and the Big Foot Beef Dip. Before or after dinner, enjoy a refreshing beverage in their cozy pub. Other activities abound as Harrison Mills is minutes away from world-class fishing, hunting, alpine skiing, paragliding, golf, and many other recreational activities.

THE BELLA COOLA PETROGLYPHS, Canada

Location: Off Highway 20 that leads into the town of Bella Coola. The entrance is a short distance southeast of the town. It may not be marked, so check with the tourist information facility in the town or inquire at the local museum. However, if you are greeted by a Native Good Will Ambassador, or are directed to one, then you can get a guided tour, which is strongly recommended.

Description: A long, winding trail that is reasonably maintained leads up into the mountains. It's a fair climb and can be slippery if the ground is wet, so the hike is not recommended for the elderly. The trail opens up into a rocky clearing where you are greeted with an array of astounding petroglyphs They were not carved by the local Bella Coola people. It is believed the area was occupied by different Native people some 10,000 years ago, who, for reasons unknown, painstakingly created the images. One image shows what certainly could be a sasquatch.

While in the town, be sure to visit Kopas Store. It was originally opened in 1937 by Cliff Kopas, a sasquatch witness, who mentions his experiences in his books that are available at the store (now run by his daughter).

Many Native people in Bella Coola area are firm believers in the existence of sasquatch. Your Good Will Ambassador will be pleased to discuss the subject with you, and might even take you to talk to some of his relatives if you ask. It's a very close community, so you don't have to travel very far.

You will see Natives net fishing on the Bella Coola River, which runs right by the town. Fishermen say that if a rock is thrown from the mountainside and splashes near their boat, they know that a sasquatch is asking for a fish. They paddle ashore and respectfully place a fish on the shoreline for their friend.

Afterword

By
Christopher L. Murphy

The sasquatch/bigfoot issue is, in my opinion, the world's most intriguing phenomenon. We have many witnesses who say they have seen the creature, too many to deny. Furthermore, the reasonably hard evidence (footprints, handprints, body prints and hair) supports the creature's existence—more support than for the existence of all people in history prior to the invention of photography. Let's face it, we believe that most people in history existed simply because someone said so.

Where the sasquatch probably lives in its greatest numbers is in the remote forest regions of Alaska, the Yukon and British Columbia. In some ways, it is easier to probe the depths of the oceans than it is these areas. The forests are so dense, and the terrain so rugged, that the only way one can explore the land is on foot. Generally speaking, only First Nations people venture into these areas—and many of them say the sasquatch exists.

Naturally, the skeptics point out that if a primate the size of the sasquatch existed, then we should have obtained firm evidence by now (i.e., bones and/or a body). I agree, we definitely should have such evidence. However, we can't simply dismiss all of the evidence we have because we are lacking "firm" evidence.

The scientific community needs to take a closer look—particularly in the areas I have mentioned. This, however, is a very tough call. It requires significant financing, and even then there are no guarantees.

Probably, the best way to see a sasquatch would be through satellite imagery (which I understand can now get close enough to read vehicle license plates). I think that if enough time were spent scouring the earth's remote regions in "real time," there would be some remarkable discoveries. If the general public were allowed access to a facility of this nature, there would be immediate results; not only regarding the sasquatch, but perhaps other creatures thought to be extinct.

Bibliography

Books

Beck, Fred, 1967 (written by Ronald Beck). *I Fought the Apemen of Mt. St. Helens.* Self published. Republished 1996, New Westminster, BC: Pyramid Publications.

Byrne, Peter, 1975. *The Search for Big Foot: Monster, Myth or Man?* Washington, DC: Acropolis Books. Republished in 1976, New York. Pocket Books.

Fusch, Ed Dr., 2002. *They Walked Among Us: Scweneyti and The Stick Indians of The Colvilles.* Riverside, Washington: Self published. www.prospectored.com/bigfoot.html

Green, John, 1978. *Sasquatch: The Apes Among Us*. Agassiz, BC: Cheam Publishing. Republished in 1981 and 2006, Surrey, BC: Hancock House Publishers Ltd.

Green, John, 2004. *The Best of Sasquatch/Bigfoot.* Surrey, BC: Hancock House Publishers Ltd.

Murphy, Christopher with John Green and Thomas Steenburg, 2004, *Meet The Sasquatch.* Surrey, BC: Hancock House Publishers Ltd.

Meldrum, Jeff, Dr., 2006. *Sasquatch: Legend Meets Science.* New York: Tom Doherty Associates (Forge book).

Mery, Fernand,1968. *The Life History and Magic of the Dog*. New York: Madison Square Press.

Morgan, Robert W., 2008. *Soul Snatchers: A Quest for the True Human Beings.* Enumclaw, Washington: Pine Winds Press, a division of Idyll Arbor Inc.

Morgan, Robert W., 2008. *The Bigfoot Pocket Field Manual.* Enumclaw, Washington: Pine Winds Press, a division of Idyll Arbor Inc.

Orchard, Vance, 2001. *The Walla Walla Bigfoot.* Prescott, Washington: Ox-Yoke Press.

Powell. Thom, 2003. *The Locals: A Contemporary Investigation of the Bigfoot/Sasquatch Phenomenon.* Surrey, BC: Hancock House Publishers Ltd.

Roosevelt, Theodore, 1893. *The Wilderness Hunter: Outdoor Pastitmes fo an American Hunter.* New York: G.P. Putnam's Sons.

Steenburg, Thomas, 1990. *Sasquatch/Bigfoot.* Surrey, BC: Hancock House Publishers Ltd.

Steenburg, Thomas, 2000. *In Search of Giants: Bigfoot/Sasquatch Encounters.* Surrey, BC: Hancock House Publishers Ltd.

Newspaper Articles

"Meat Eaters They Appear to Be!," 1969. Stevenson, Washington: *Skamania County Pioneer.* March 25.

"Monkey Like Wildman," 1904. Vancouver, BC:*Daily Province.*

"The Seeahtik Belief—Clues to Gorilla Men Found, May Be Lost Race of Giants," 1924. Seattle, Washington: *Seattle Daily Times.* July 16.

"What is it? A Strange Creature Captured Above Yale: A British Columbia Gorilla," 1884. Victoria, BC: *The Colonist.* July 3. (Note: Actual article is incorrectly dated as 1882.)

Other Publications/Papers

Crowe, Ray. *The Track Record* (newsletter). Hillsboro, Oregon.

Fahrenbach, Henner, Dr., 1998. "Sasquatch: Size, Scaling, and Statistics." *Cryptozoology,* Vol. 13:47–75.

Fahrenbach, Henner, Dr., 1997. "Sasquatch Smell, Aroma, Odor, Scent." www.bigfootencounters.com/biology/smell.htm

Photographs/Illustrations — Credits/Copyrights

Page	Item	Credit
	Front cover scene	C. Murphy
	Front/back cover eye	C. Murphy
18	Gram with beets	L. Coil Suchy
22	Cake	L. Coil Suchy
23	House	L. Coil Suchy
23	Clothsline	L. Coil Suchy
23	Gram with squash	L. Coil Suchy
24	Gram with cabbages	L. Coil Suchy
24	Gram with assorted vegetables	L. Coil Suchy
27	Stamp	Canada Post Corp. 1957
31	Jacko on track	D. Hopkins
31	Jacko in boxcar	D. Hopkins
33	Men, torches, sasquatch	G. Krejci
36	T. Roosevelt	Public domain
36	Book cover	Public domain
41	Drawing of sasquatch head	I. Sanderson
42	Ostman and Green	J. Green
42	Sworn statement	J. Green
44	Beck with rifle	J. Green
44	Group at cabin	Public domain
44	Beck and Smith	Public domain
44	Ape Canyon	J. Green
45	Peter Byrne	P. Byrne
46	River/inlet scene	P. Byrne
47	Map	P. Byrne
48	Father Terhaar	P. Byrne
50	Ruby Creek house	J. Green
51	R. Dahinden	J. Green
52	Mica Mountain	C. Murphy
53	Drawing	C. Murphy
55	Affidavit	J. Green
72	Lummi scene	C. Murphy
72	Totem pole, left	C. Murphy
72	Totem pole, right	C. Murphy
81	Bill White	B. White
83	Plane	B. White

172	Drawing	F. O'Connor
213	Dr. Jeff Meldrum	J. Meldrum
217	Book cover	J. Meldrum/T.Doherty Associates
218	Dr. H. Fahrenbach	H. Fahrenbach
224	Bob Gimlin	C. Murphy
229	Group	C. Murphy
230	John Pickering	J. Pickering
237	Ray Crowe	R. Crowe
242	Sean Fries	S. Fries
248	Michael Rugg	M. Rugg
252	"The Moment"	M. Rugg
253	Ray Rosa	R. Rosa
256	Robert Morgan	R. Morgan
264	Book cover *Soul Snatchers*	R. Morgan/Pine Winds Press
264	Book cover *Field Manual*	R. Morgan/Pine Winds Press
265	Tom Biscardi	T. Biscardi
272	Group	T. Biscardi
273	Thomas Steenburg	C. Murphy
277	Book cover *In Search...*	T. Steenburg/Hancock House
277	Book cover *Sasquatch...*	T. Steenburg/Hancock House
278	John Green	K. Walls
282	Book cover *Sasquatch...*	J. Green/Hancock House
282	Book cover *Best of...*	J. Green/Hancock House
283	Dmitri Bayanov	D. Bayanov
290	Book cover *Americas BF*	D. Bayanov/Crypto Logos
290	Book cover *In the Footsteps*	D. Bayanov/Crypto Logos
291	Peter Byrne	P. Byrne
297	Book Cover *Big Foot...*	P. Byrne/Pocket Books
297	Bigfoot Information Center	P. Byrne
297	Logo	Boston Academy of Applied Scie
305	Frame 352, P/G Film	R. Patterson/Public Domain
306	Casts, film site	C. Murphy
307	Film site model	C. Murphy
308	Footprint in ground	R. Titmus
308	Bob Titmus	R. Titmus
309	Casts Cripplefoot	C. Murphy
309	Casts with bones	C. Murphy
310	Hand cast (top)	C. Murphy
310	Hand cast (lower)	C. Murphy

311	Hand cast	C. Murphy
312	Giganto skull	C. Murphy
312	Three skulls	C. Murphy
313	Hair in container	C. Murphy
313	Hair strand	H. Fahrenbach
314	Heel cast	C. Murphy
314	Human foot	C. Murphy
315	Dermal ridge chart	C. Murphy
316	Drawing	C. Rupp
317	Stone foot (all photos)	C. Murphy
318	Chehalis mask	C. Murphy
319	Kwakiutl mask	C. Murphy
320	Haida mask	C. Murphy
351	Map	C. Murphy/Y. Leclerc
353	Stone head	C. Murphy
353	Petroglyph	K. Moskowitz-Strain
354	Pictograph	K. Moskowitz-Strain
354	Drawing	B. Bannon/K. Moskowitz Strain
354	Almanac drawing	B. Regal
355	Woman poaching acorns	Public domain
355	Stonish Giants	Public domain
356	Mask drawing	P. Travers
356	Footprint	C. Murphy collection
357	Footprint (top)	R. Lyle Laverty
357	Footprint (lower)	R. Lyle Laverty
358	Cast footprint	C. Murphy
359	Hand drawing	B. Jenkins/C. Murphy
360	Creature running	Z. Hamilton
361	Sasquatch portrait	C. Murphy
362	Sasquatch portrait	P. Travers
363	Sasquatch heads (4)	C. Murphy
364	Sculpture	C. Murphy
365	Frame 352, P/G film	R. Patterson/Public domain
365	Wood fragment	C. Murphy
366	Stamp	Canada Post Corp.
366	Bob Titmus	J. Green
367	Totem pole	J. Green
368	Kwakiutl mask (top)	D. Hancock
368	Kwakiutl mask (lower)	D. Hancock

General Index

hancock
house